Matthias Preisig

Modeling two-phase flows on moving domains

Matthias Preisig

Modeling two-phase flows on moving domains

Südwestdeutscher Verlag für Hochschulschriften

Impressum/Imprint (nur für Deutschland/ only for Germany)
Bibliografische Information der Deutschen Nationalbibliothek: Die Deutsche Nationalbibliothek
verzeichnet diese Publikation in der Deutschen Nationalbibliografie; detaillierte bibliografische
Daten sind im Internet über http://dnb.d-nb.de abrufbar.
Alle in diesem Buch genannten Marken und Produktnamen unterliegen warenzeichen-, marken-
oder patentrechtlichem Schutz bzw. sind Warenzeichen oder eingetragene Warenzeichen der
jeweiligen Inhaber. Die Wiedergabe von Marken, Produktnamen, Gebrauchsnamen,
Handelsnamen, Warenbezeichnungen u.s.w. in diesem Werk berechtigt auch ohne besondere
Kennzeichnung nicht zu der Annahme, dass solche Namen im Sinne der Warenzeichen- und
Markenschutzgesetzgebung als frei zu betrachten wären und daher von jedermann benutzt
werden dürften.

Verlag: Südwestdeutscher Verlag für Hochschulschriften Aktiengesellschaft & Co. KG
Dudweiler Landstr. 99, 66123 Saarbrücken, Deutschland
Telefon +49 681 37 20 271-1, Telefax +49 681 37 20 271-0, Email: info@svh-verlag.de
Zugl.: Lausanne, EPFL, Dissertation, 2008

Herstellung in Deutschland:
Schaltungsdienst Lange o.H.G., Berlin
Books on Demand GmbH, Norderstedt
Reha GmbH, Saarbrücken
Amazon Distribution GmbH, Leipzig
ISBN: 978-3-8381-0373-0

Imprint (only for USA, GB)
Bibliographic information published by the Deutsche Nationalbibliothek: The Deutsche
Nationalbibliothek lists this publication in the Deutsche Nationalbibliografie; detailed
bibliographic data are available in the Internet at http://dnb.d-nb.de.
Any brand names and product names mentioned in this book are subject to trademark, brand
or patent protection and are trademarks or registered trademarks of their respective holders.
The use of brand names, product names, common names, trade names, product descriptions
etc. even without a particular marking in this works is in no way to be construed to mean that
such names may be regarded as unrestricted in respect of trademark and brand protection
legislation and could thus be used by anyone.

Publisher:
Südwestdeutscher Verlag für Hochschulschriften Aktiengesellschaft & Co. KG
Dudweiler Landstr. 99, 66123 Saarbrücken, Germany
Phone +49 681 37 20 271-1, Fax +49 681 37 20 271-0, Email: info@svh-verlag.de

Copyright © 2009 by the author and Südwestdeutscher Verlag für Hochschulschriften
Aktiengesellschaft & Co. KG and licensors
All rights reserved. Saarbrücken 2009

Printed in the U.S.A.
Printed in the U.K. by (see last page)
ISBN: 978-3-8381-0373-0

Acknowledgments

This thesis would not have been possible without the interaction with several people, to whom I wish to express my gratitude.

Most of all I would like to thank my adviser, Prof. Thomas Zimmermann, for his guidance, suggestions and his never ending flow of ideas. This thesis is the result of a great number of intense and sometimes heated discussions, at the end of which there almost always stood a tangible result.

The financial support by the Swiss National Science Foundation under grants no. 2100-067954 and no. 200020-113295 is gratefully acknowledged.

I wish to thank Prof. Dominique Eyheramendy, Prof. Boris Jeremic and Dr Marco Picasso for reviewing my thesis thoroughly and for giving valuable comments that helped to improve the quality of the document considerably. I also wish to thank Prof. Eugen Brühwiler for having accepted to act as president of the jury.

Many thanks go to my former colleagues of the Structural and Continuum Mechanics Laboratory as well as to my current colleagues of the Computational Solid Mechanics Laboratory, for being part of an environment in which it is a pleasure to work, and for helping with many small and big problems. In particular the programming skills and an implementation of an algorithm for spatial search provided by Guillaume Anciaux greatly increased the speed and memory efficiency of my research code.

Also I would like to thank Thomas Hughes for letting me spend five months in his group at University of Texas at Austin. This visit allowed me to discuss my work with other people and to get a first-hand look into many exciting new fields of research.

Finally, I am very grateful to my parents, my sister, and my friends for their support and occasional distractions from work, without which this work would never have been possible.

Contents

1 Introduction — 1

2 Updated Lagrangian modeling of incompressible free-surface single-phase flows — 7
  2.1 Introduction — 8
  2.2 Updated Lagrangian formulation — 8
  2.3 Governing equations and weak form — 9
     2.3.1 Mass conservation — 9
     2.3.2 Momentum conservation — 10
     2.3.3 Constitutive relation — 11
     2.3.4 Summary of the initial/boundary value problem — 12
     2.3.5 Weak form — 12
     2.3.6 Semidiscrete matrix form — 13
  2.4 Time integration scheme — 13
     2.4.1 Generalized trapezoidal family of methods — 13
     2.4.2 Mesh update — 16
  2.5 Spatial discretization and numerical approximation — 16
     2.5.1 Numerical approximation — 18
     2.5.2 Mesh-independent finite element method — 25
     2.5.3 Volumetric locking and stabilization of the mixed formulation — 27
     2.5.4 Re-meshing — 30
     2.5.5 Re-mapping procedure — 31
     2.5.6 Detection of contact at the fluid boundary — 31
  2.6 Numerical tests — 32
     2.6.1 Hydrostatic patch test — 32
     2.6.2 Transient lid-driven cavity flow — 34
     2.6.3 Stratified flow — 36
     2.6.4 Flow over a backward-facing step — 39

	2.6.5	Flow over a backward-facing step with free surface	42
	2.6.6	Solitary wave	45
	2.6.7	Formation of drop due to surface tension effect	47
	2.6.8	Dam break	51
2.7	Conclusions		53

3 Updated Lagrangian method for modeling of multi-phase free-surface flows 57
- 3.1 Introduction 58
  - 3.1.1 Classification of mud- and debris-flow models 58
  - 3.1.2 Review of mud- and debris-flow literature 59
  - 3.1.3 Proposed model 60
- 3.2 Governing equations 62
  - 3.2.1 Volume averaging of governing equations of two-phase flow 63
  - 3.2.2 Conservation of mass 63
  - 3.2.3 Conservation of momentum 64
  - 3.2.4 Momentum exchange 65
  - 3.2.5 Constitutive relation 66
  - 3.2.6 Summary of the initial/boundary value problem 67
- 3.3 Weak form and stabilization 68
  - 3.3.1 Stabilization 69
  - 3.3.2 Semidiscrete matrix form 70
- 3.4 Time integration scheme 71
  - 3.4.1 Mesh update 71
- 3.5 Spatial discretization 74
- 3.6 Computation of volume fractions 75
  - 3.6.1 Background 75
  - 3.6.2 Current Method 77
- 3.7 Numerical tests 83
  - 3.7.1 Equivalence between two-phase and single-phase fluid 83
  - 3.7.2 Sedimentation 84
  - 3.7.3 Flow over a backward-facing step with free surface 89
  - 3.7.4 Sharp gradients of volume fractions 91
  - 3.7.5 Dam break 94
  - 3.7.6 Mudflow impacting an obstacle 94
- 3.8 Conclusions 98

Contents

4 Implementation   **101**
- 4.1 Introduction 102
  - 4.1.1 Object-oriented finite element programming 102
  - 4.1.2 An object-oriented finite element modeling on moving domains 102
- 4.2 Organization of the research code 104
  - 4.2.1 `Analysis` class 106
  - 4.2.2 `Discretization` class 106
  - 4.2.3 `Mesh` class 108
  - 4.2.4 Computation of elemental matrices and arrays 110
  - 4.2.5 Assembly of RHS and LHS 111
  - 4.2.6 Other additions 111
- 4.3 Spatial searching 112
- 4.4 CPU-time 113

5 Concluding remarks   **115**
- 5.1 Conclusions 116
- 5.2 Further research 116
  - 5.2.1 Constitutive modeling 116
  - 5.2.2 Pseudo three-phase formulation 116
  - 5.2.3 Fluid-structure interaction 117
  - 5.2.4 Remarks for extension to three spatial dimensions 118

A Appendix   **119**
- A.1 Elemental matrices and arrays for single-phase flow 120
- A.2 Delaunay triangulation and α-shapes 120
  - A.2.1 Constrained Delaunay triangulation and Delaunay mesher 121
- A.3 Elemental matrices and arrays for two-phase flow 122
- A.4 Computing volume fractions using mass conservation of each phase 125

References   **127**

6

Chapter 1

Introduction

Every year throughout the world, debris flows cause an immense amount of damage to property and people. The growth of population inevitably puts a lot of pressure on developers to build houses in areas that are exposed to elevated risk of debris flows, mudflows, landslides and similar events which are often of a hydro-geological nature. This risk is accentuated by high demand for real estate located in topographically attractive places. Mitigation of the risk is two-fold: One, mitigation requires better zoning and urban planning measures to single out danger zones. Two, it requires a better understanding of these events in order to protect the existing infrastructure from damage. In both cases, numerical modeling can play an important role in risk mitigation by one, helping to establish maps that outline areas which are in the flow path, and two, computing characteristics of the event such as depth of the flow or slide and forces, both necessary for designing effective barriers, walls, nets or other protection devices.

Debris flows are flows of a mixture of water with soil, rocks, trees and other debris. They are in general initiated by a slope failure which can be triggered by, for example, rainfall, strong groundwater flow or snow melt. Alternatively, a debris flow can also be caused by a sudden surge of water, caused by the breaking of a dam, rockfall or a landslide into a lake causing the water to overflow. As the flood surge rushes down a dry channel it can transform into a debris flow by eroding the bed and picking up material on its way. The area affected by a debris flow can be separated into an initiation zone from where most of the debris originates, a propagation zone, usually a narrow channel or canyon, and a deposition zone where the solid component of the flow is deposited due to a slope change and a decrease in velocity as a result thereof.

From a modeling point of view, debris flows are characterized as gravity-driven flows of a multiphase material. During the event, the material undergoes large motion, which causes the boundary to change continuously. The boundary, and in particular the free surface, is not known a priori, but part of the solution.

Current state-of-the-art numerical models of debris flow use the thin-layer assumption, stating that the extension of the flow in and perpendicular to the direction of flow is much larger than the depth of the flow. This justifies the application of depth-averaging, where the flow is projected onto a line for simple channel models or onto a two-dimensional mesh (a map). The depth of the flow and the velocity are part of the solution in each mesh node. A hypothesis on the vertical distribution of stresses is required to integrate the governing equations over depth.

Numerical modeling of debris flows with depth-averaged models has been a subject of active research for many years and is still ongoing. Such models have proven to be very valuable for evaluating run-out distances and flow paths in order to edit hazard maps. Specialized consulting firms have acquired extensive knowledge in applying existing models. Due to the thin-layer assumption, however, the stresses cannot be predicted with very high accuracy and the application of these models for obtaining loads acting on a protection device is very limited.

As an alternative to numerical simulation, in situ measurements in instrumented debris-flow channels can give results that can be extrapolated to other sites. Full-scale experiments on test sites, where a debris mixture is released onto an instrumented prototype of a protection system, are another direction of current debris-flow research. Although, considering the high cost of such experiments, it is obvious that numerical simulation can play a more important role in the development of debris flow protection systems. In order to obtain results that have the required degree of reliability, the next generation numerical model has to be full scale three dimensional. The present work is an important step in this direction.

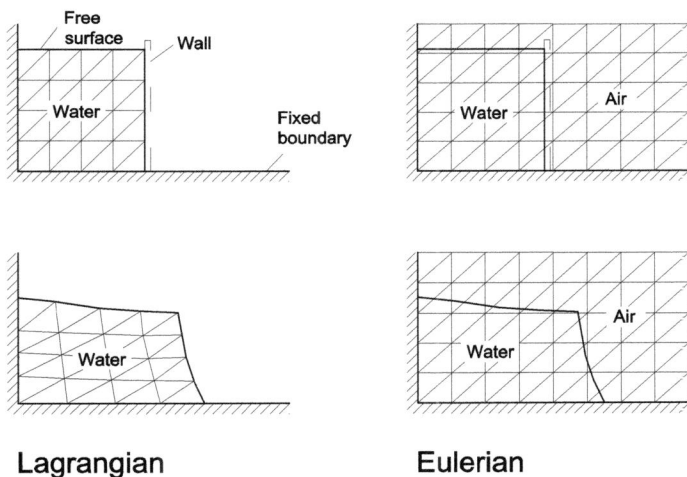

Figure 1.1: Simulation of the dam-break problem with a Lagrangian and an Eulerian mesh.

The goal of this work is to be able to simulate the impact of a mud- or debris flow, that is running down a slope, on an obstacle, and to obtain detailed time histories of forces acting on the obstacle. The model has to be able to take into account the complex material behavior of a debris-water mixture. Focus is put on the development of a robust algorithmic framework for describing the motion of the mixture. The claim is that the main features of multi-phase flow can be captured by algorithmically coupling two materials that each follow a very simple constitutive model. For the sake of simplicity, we restrict ourselves to two dimensions in the development of the model. However, most of the components of the model remain valid in three dimensions without any major modifications.

Full scale modeling of a fluid undergoing large motion and whose boundary changes continuously poses special requirements to the spatial description. We chose a formulation where particles move according to their mass and acceleration while being subject to forces from the surrounding fluid. The incompressible Navier-Stokes equations are solved in a Lagrangian reference frame at each time step. The Lagrangian reference frame is updated (moved) to the current spatial position after each step. Since the Lagrangian frame of reference is attached to the material, no discretization is required in zones where no material is present. This is in contrast to Eulerian methods, where, in general, large parts containing air have to be modeled. Figure 1.1 illustrates the meshes used in an Eulerian and in a Lagrangian simulation of a so-called dam-break problem. The figure shows the outline of the water-filled domain at the beginning of the simulation and shortly after the lateral wall, representing the dam, has been removed. The free surface is part of the solution in the Lagrangian case. In the Eulerian case the position of the free surface has to be computed separately, for example by using a level-set technique or a surface-reconstruction algorithm. Such a computation can be costly and introduces additional approximations.

By formulating the problem in a moving reference frame another important disadvantage of Eulerian methods is avoided. The total derivative of the velocity gives rise to a convective term which is not present in a Lagrangian formulation. On one hand, such a convective term causes spurious oscillations in the velocity field, unless the equations are stabilized. On the other hand, Lagrangian methods suffer from problems related to distortion of the mesh. This can be dealt with by re-zoning the nodes and reconnecting the mesh. The use of meshless methods is also an alternative. In the framework of simulating free-surface flows it is, however, our opinion that the advantages of Lagrangian meshes outweigh their disadvantages.

The material behavior of a mudflow is governed by the properties of its constituents, water and granular material. The grain sizes of the granular material range from very fine particles, as in clay, to gravel and even boulders. We restrict ourselves in this work to material with an important fine content, a characteristic property of sediments deposited by the melting of a glacier. The high fine content gives the mixture a muddy, viscous behavior. Keeping in mind that the algorithmic framework is not specific to any particular kind of material, we limit the scope of this work to a two-phase material where both the fluid and the solid phase are modeled as viscous fluids.

Both phases are smeared over the fluid domain. This means that no phase interfaces are considered. The presence of either phase is specified by its volume fraction. Figure 1.2a) illustrates the reality whereas Figure 1.2b) shows the distribution of smeared solid volume fractions.

The range of Reynolds numbers that the proposed method is expected to cover is below $Re = 1000$. This covers the laminar flow regime and can be considered a reasonable assumption

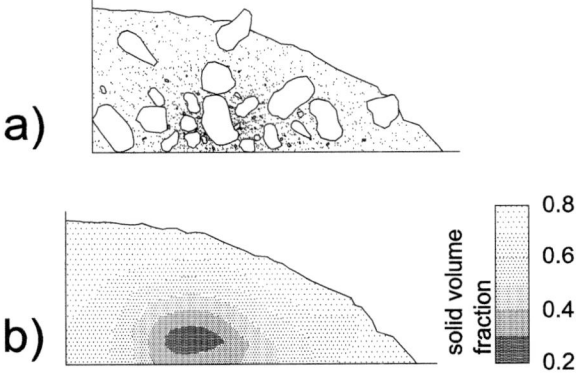

Figure 1.2: The mass of the solid phase is smeared over the mixture.

for mudflows with high contents of fine particles. In order to lift this limitation, a turbulence model would have to be included.

The main difficulty in developing an updated Lagrangian method for simulating a two-phase fluid lies in finding an appropriate way to update the spatial coordinates of the nodes. Since in each node the velocities of two phases are computed, the update produces two new nodes, one for each phase. Multiplying the number of nodes by two at each Lagrangian update is not a solution for obvious reasons. We deal with this fundamental problem by re-creating a new set of nodes, on which the solution transported by the nodes at their updated spatial coordinates is interpolated. This approach solves at the same time the problem of mesh distortion.

The original features of the presented method are a constitutive model of a two-phase fluid that we derive from a single-phase viscous fluid, an algorithm to perform the Lagrangian update for a two-phase material, and an innovative method for computing volume fractions of both phases based on the spatial coordinates of the nodes before and after the Lagrangian update. This work also sheds a new light on updated Lagrangian modeling of single-phase flow using meshless methods and it shows that most of the advantages of meshless methods can be obtained by combining finite elements with a re-meshing strategy.

The thesis is structured as follows: Chapter 2 establishes the fundamentals of updated Lagrangian modeling of single-phase flows with a free surface. The spatial discretization technique is presented and the choice of finite elements over meshless methods is justified. A series of numerical examples validate the method and show that frequent re-meshing doesn't deteriorate the results. In Chapter 3, the method for modeling two-phase flow is laid out. The algorithm for updating the spatial coordinates of the nodes considering two different phase velocities and the re-meshing and re-mapping procedures are given. Numerical tests validate and show the versatility of the method. Chapter 4 gives a detailed description of the implementation of the method into a computer code. Finally, in Chapter 5 we conclude and present a few ideas for further research.

Chapter 2

Updated Lagrangian modeling of incompressible free-surface single-phase flows

2.1 Introduction

Accurate capturing of the free surface is essential in simulations of fluid flow problems, in which the position of the fluid boundary is not known a priori. Arbitrary Lagrangian-Eulerian methods (ALE) have been used extensively in the past (see, for example [26] or [5] for some more recent developments). ALE methods in general use a mesh that coincides with the free-surface while interior nodes are moved independently of the fluid. Purely Eulerian methods on the other hand use a mesh that is fixed in space. Because the free surface in general doesn't coincide with element edges it has to be defined in some other way. The Marker-and-Cell method (MAC) is one of the earliest developments in this direction, the Volume-of-Fluid method (VOF) and level-set methods are others. The abundance of different methods indicates that all these methods have their strength and weaknesses. Some conserve mass with high accuracy, some are conceptually very appealing and others are computationally efficient. None of these methods seems to have all of these mentioned properties. One disadvantage is shared by all Eulerian and ALE methods: The presence of a convective term that requires stabilization in order to avoid spurious oscillations.

A more elegant way to model free surface flows consists in using Lagrangian meshes. For examples where updated Lagrangian discretizations have been used for simulating free-surface flows see [35, 36, 21]. Conceptually speaking we track particles of fluid that transport their mass and acceleration. These particles are subject to viscous stresses due to the surrounding fluid. This point of view is similar to the one adopted in Idelsohn et al. [31].

In a Lagrangian discretization nodes move with the material. Only the region occupied by the fluid has to be modeled. The definition of the free surface is given naturally by the position of the nodes that lie on it. No convective term requiring stabilization appears in the momentum equation.

2.2 Updated Lagrangian formulation

In a Lagrangian formulation the position of a node is given by its material coordinate $\mathbf{X}$, that is the initial position of the material point at time $t = 0$. The spatial coordinate $\mathbf{x}$ of that same node at time t_{n+1} defines the motion ϕ

$$\mathbf{x} = \phi(\mathbf{X}, t_{n+1}) \qquad (2.2.1)$$

The displacement of the node with material coordinate $\mathbf{X}$ is defined as the difference between the current and the initial coordinate

$$\mathbf{d}(\mathbf{X}, t_{n+1}) = \phi(\mathbf{X}, t_{n+1}) - \phi(\mathbf{X}, t = 0) = \phi(\mathbf{X}, t_{n+1}) - \mathbf{X} \qquad (2.2.2)$$

2.3 – Governing equations and weak form

In the updated Lagrangian formulation all variables are referred to the last known configuration. The displacement increment of a node $\mathbf{X}$ is defined as the difference between the current coordinate and the coordinate at the last step

$$\Delta \mathbf{d}(\mathbf{X}, t_{n+1}) = \phi(\mathbf{X}, t_{n+1}) - \phi(\mathbf{X}, t_n) = \phi(\mathbf{X}, t_{n+1}) - \mathbf{X} \tag{2.2.3}$$

Integrals are evaluated on the last known configuration, and derivatives are expressed with respect to the last known coordinates. The term 'updated' stems from the fact that the domain over which the integrals of the weak form are evaluated has to be updated at each time step or iteration.

2.3 Governing equations and weak form

We derive the governing equations from conservation of mass and momentum, evaluated on a volume V_x.

2.3.1 Mass conservation

The material derivative $\frac{D}{Dt}$ of the mass contained in a volume V_x is equal to zero.

$$\frac{D}{Dt} \int_{V_x} \rho \, dV_x = 0 \tag{2.3.1}$$

ρ is the mass density. In order to obtain a differential equation, which is valid pointwise, we first have to move the derivative inside the integral. To do this the volume V_x has to be expressed with respect to its material configuration V_X.

$$\begin{aligned}
\frac{D}{Dt} \int_{V_x} \rho \, dV_x &= \frac{D}{Dt} \int_{V_X} \rho J \, dV_X \\
&= \int_{V_X} \frac{D \rho J}{Dt} \, dV_X
\end{aligned} \tag{2.3.2}$$

Now we apply Reynold's transport theorem and transform the integral back to the spatial frame of reference.

$$\begin{aligned}
\int_{V_X} \frac{D \rho J}{Dt} \, dV_X &= \int_{V_X} \left(\frac{D \rho}{Dt} J + \rho \frac{D J}{Dt} \right) dV_X \\
&= \int_{V_X} \left(\frac{D \rho}{Dt} + \rho \nabla \cdot \mathbf{v} \right) J \, dV_X \\
&= \int_{V_x} \left(\frac{D \rho}{Dt} + \rho \nabla \cdot \mathbf{v} \right) dV_x
\end{aligned} \tag{2.3.3}$$

Here we used the fact that $\frac{DJ}{Dt} = \frac{D}{Dt}\frac{\partial x}{\partial X} = J\nabla \cdot \mathbf{v}$, where $\mathbf{v}$ is the velocity of a material point. Arbitrariness of the volume V_x allows us to write

$$\frac{D\rho}{Dt} + \rho \nabla \cdot \mathbf{v} = 0 \qquad (2.3.4)$$

or, alternatively

$$\frac{\partial \rho}{\partial t} + \nabla \cdot (\rho \mathbf{v}) = 0 \qquad (2.3.5)$$

If the fluid is incompressible, then the mass density is constant ($\frac{D\rho}{Dt} = 0$). In this case the equation of mass conservation reduces to

$$\nabla \cdot \mathbf{v} = 0 \qquad (2.3.6)$$

2.3.2 Momentum conservation

The rate of change of the momentum of a volume V_x is equal to the stresses σ acting on the boundary S_x of the volume V_x and the body force $\mathbf{b}$

$$\frac{D}{Dt}\int_{V_x} \rho \mathbf{v} \, dV_x = \int_{S_x} \sigma \cdot \mathbf{n} \, dS_x + \int_{V_x} \mathbf{b} \, dV_x \qquad (2.3.7)$$

Again, we need to transfer the term on the left-hand side to the material configuration and apply Reynold's transport theorem

$$\begin{aligned}
\frac{D}{Dt}\int_{V_x} \rho \mathbf{v} \, dV_x &= \frac{D}{Dt}\int_{V_X} \rho \mathbf{v} J \, dV_X \\
&= \int_{V_X} \frac{D\rho \mathbf{v} J}{Dt} \, dV_X \\
&= \int_{V_X} \left(\frac{D\rho}{Dt}\mathbf{v} J + \frac{D\mathbf{v}}{Dt}\rho J + \frac{DJ}{Dt}\rho \mathbf{v}\right) dV_X \\
&= \int_{V_X} \left(\frac{D\rho}{Dt}\mathbf{v} + \frac{D\mathbf{v}}{Dt}\rho + \rho \mathbf{v}\nabla \cdot \mathbf{v}\right) J \, dV_X \qquad (2.3.8)
\end{aligned}$$

Now we can use the equation of mass conservation (Equation 2.3.4) and transfer the integral back to the spatial domain

$$\begin{aligned}
\int_{V_X} \left(\frac{D\rho}{Dt}\mathbf{v} + \frac{D\mathbf{v}}{Dt}\rho + \rho \mathbf{v}\nabla \cdot \mathbf{v}\right) J \, dV_X &= \int_{V_X} \left(\frac{D\mathbf{v}}{Dt}\rho\right) J \, dV_X \\
&= \int_{V_x} \left(\frac{D\mathbf{v}}{Dt}\rho\right) dV_x \qquad (2.3.9)
\end{aligned}$$

2.3 – Governing equations and weak form

Using the divergence theorem on the first term on the right hand side of Equation 2.3.7 we obtain

$$\int_{S_x} \sigma \cdot \mathbf{n} \, dS_x = \int_{V_x} \nabla \cdot \sigma \, dV_x \qquad (2.3.10)$$

Substituting Equations 2.3.9 and 2.3.10 into Equation 2.3.7 yields

$$\int_{V_x} \left(\frac{D\mathbf{v}}{Dt} \rho \right) dV_x = \int_{V_x} \nabla \cdot \sigma \, dV_x + \int_{V_x} \mathbf{b} \, dV_x \qquad (2.3.11)$$

Because this has to be true for any control volume V_x the following holds:

$$\rho \frac{D\mathbf{v}}{Dt} = \nabla \cdot \sigma + \mathbf{b} \qquad (2.3.12)$$

2.3.3 Constitutive relation

The stress tensor σ can be decomposed into a viscous stress tensor τ and a pressure p.

$$\sigma = \tau + p\mathbf{I} \qquad (2.3.13)$$

Remark 2.3.1 *We adopt the following sign convention: Applying a positive pressure to a compressible medium causes the medium to expand.*

For a Newtonian fluid with a dynamic viscosity of μ the viscous stress tensor is

$$\tau = \mu(\nabla \mathbf{v} + (\nabla \mathbf{v})^T) + \lambda(\nabla \cdot \mathbf{v})\mathbf{I} \qquad (2.3.14)$$

Remark 2.3.2 *In order to guarantee that the normal stress acting on a surface in a fluid at rest is equal to the pressure p, the viscous stress tensor τ has to be purely deviatoric. This leads to*

$$\text{tr}\,\tau = (2\mu + 3\lambda)\nabla \cdot \mathbf{v} = 0$$

Substituting $\lambda = -2/3\mu$ into Equation 2.3.14 yields the following expression for τ:

$$\tau = \mu \left[\nabla \mathbf{v} + (\nabla \mathbf{v})^T - \frac{2}{3}(\nabla \cdot \mathbf{v})\mathbf{I} \right]$$

Substituting the strain rate tensor $\dot{\epsilon} = \frac{1}{2}(\nabla \mathbf{v} + (\nabla \mathbf{v})^T)$ into the constitutive relation yields:

$$\tau = 2\mu \left(\dot{\epsilon} - \frac{1}{3}(\nabla \cdot \mathbf{v})\mathbf{I} \right) \qquad (2.3.15)$$

Remark 2.3.3 *As we have seen in Equation 2.3.6 the divergence of the velocity is equal to zero for incompressible materials. In order to remain most general we leave it in the formulation. In the next chapter, which treats two-phase flow, the divergence of the phase velocities is not zero anymore.*

2.3.4 Summary of the initial/boundary value problem

Now we can summarize the boundary value problem on a domain Ω and for $]0,T[$, an open time interval of length T. For incompressible Newtonian fluids we state: Given $\mathbf{b}$, $\mathbf{g}$, $\mathbf{h}$, $\mathbf{v}_0$ and p_0, find $\mathbf{v}$ and p on $\Omega \times]0,T[$ such that

$$\rho \frac{D\mathbf{v}}{Dt} = \nabla \cdot \tau + \nabla p + \mathbf{b} \quad \text{on } \Omega \times]0,T[\quad (2.3.16)$$

$$\nabla \cdot \mathbf{v} = 0 \quad \text{on } \Omega \times]0,T[\quad (2.3.17)$$

$$\mathbf{v} = \mathbf{g} \quad \text{on } \partial\Omega_g \times]0,T[\quad (2.3.18)$$

$$\sigma \cdot \mathbf{n} = \mathbf{h} \quad \text{on } \partial\Omega_h \times]0,T[\quad (2.3.19)$$

$$\mathbf{v}(t=0) = \mathbf{v}_0 \quad \text{on } \Omega \quad (2.3.20)$$

$\partial\Omega_g$ denotes the Dirichlet part of the boundary of Ω, the part on which we impose the displacement $\mathbf{g}$, and $\partial\Omega_h$ the Neumann part, where surface tractions $\mathbf{h}$ are imposed.

2.3.5 Weak form

Let $\mathcal{S}_i = \{v_i \in H^1(\Omega) \mid v_i = g_i \text{ on } \Gamma_{g_i}\}$ be a space of trial functions, $\mathcal{V}_i = \{w_i \in H^1(\Omega) \mid w_i = 0 \text{ on } \Gamma_{g_i}\}$ a space of test functions and $\mathcal{P} = \{p \in L^2(\Omega)\}$ a space of both trial and test functions [1]. Then the weak form of Equations 2.3.16 to 2.3.20 can be stated as: Find $v_i(t) \in \mathcal{S}_i$ and $p(t) \in \mathcal{P}$, $t \in [0,T]$, such that for all $w_i \in \mathcal{V}_i$ and $q \in \mathcal{P}$ the following equation holds:

$$B(\mathbf{w}, q; \mathbf{v}, p) = L(\mathbf{w}, q) \quad (2.3.21)$$

where

$$B(\mathbf{w}, q; \mathbf{v}, p) = \int_\Omega \rho \mathbf{w} \cdot \frac{D\mathbf{v}}{Dt} d\Omega + \int_\Omega \nabla \mathbf{w} : \mathbf{D} : \nabla \mathbf{v} d\Omega + \int_\Omega \nabla \cdot \mathbf{w} p d\Omega$$
$$+ \int_\Omega q \nabla \cdot \mathbf{v} d\Omega \quad (2.3.22)$$

$$L(\mathbf{w}, q) = \int_\Omega \rho \mathbf{w} \cdot \mathbf{b} d\Omega + \int_{\Gamma_h} \mathbf{w} \cdot \mathbf{h} d\Gamma - \int_\Omega \nabla \mathbf{w} : \mathbf{D} : \nabla \mathbf{g} d\Omega$$
$$- \int_\Omega q \nabla \cdot \mathbf{g} d\Omega \quad (2.3.23)$$

$\mathbf{D}$ is the constitutive matrix, defined in Appendix A.1. The viscous stress tensor can be written as $\tau = \mathbf{D} : \nabla \mathbf{v}$.

---

[1] $H^1(\Omega)$ is the space of functions u for which the inner product $(u,u)_1 = \int_\Omega (u^2 + (u_{,i})^2) d\Omega$ is finite and the norm $||u||_1 = \sqrt{(u,u)}$ exists (i denotes the spatial dimensions. $i = 1, 2$ for 2D and $i = 1, 2, 3$ for 3D). $L_2(\Omega)$ is the space of functions u for which the inner product $(u,u) = \int_\Omega u^2 d\Omega$ is finite and the norm $||u||_{L_2} = \sqrt{(u,u)}$ exists.

2.4 – Time integration scheme

Note that in the equations of the weak form no frame of reference has yet been specified. The integrals in Equations 2.3.22 and 2.3.23 can be evaluated on a fixed spatial domain $\Omega(x)$ or on a moving domain $\Omega(X)$ attached to the material. This choice is dictated by what is more convenient in evaluating the integrals. As mentioned earlier we choose an updated Lagrangian formulation, where the integrals and spatial derivatives are evaluated on the last known configuration $\Omega(X, t_n)$.

2.3.6 Semidiscrete matrix form

Introducing a set of approximations discretizing the space into Equation 2.3.21 and using arbitrariness of the test functions leads to a semidiscrete matrix equation where time is kept continuous

$$\mathbf{Ma} + \mathbf{Kv} = \mathbf{f} \tag{2.3.24}$$

a is a vector of nodal accelerations and **v** is a vector of nodal velocities and pressures:

$$\mathbf{a} = [a_x^1 \ a_y^1 \ 0 \ \cdots \ a_x^I \ a_y^I \ 0 \ \cdots \ a_x^n \ a_y^n \ 0]^T$$
$$\mathbf{v} = [v_x^1 \ v_y^1 \ p^1 \ \cdots \ v_x^I \ v_y^I \ p^I \ \cdots \ v_x^n \ v_y^n \ p^n]^T$$

where the superscript $I \in \{1, \cdots, n\}$ identifies a node. We now proceed by presenting a time-stepping algorithm that solves Equation 2.3.24 at a time $t = t_{n+1}$. Spatial discretization is described in Section 2.5.

2.4 Time integration scheme

Equation 2.3.24 is non-linear since both matrices **M** and **K** change at each time the nodes of the mesh are moved. The unknowns of the equation are velocity and acceleration. The displacement is required to update the nodal coordinates at the end of each iteration, but it is not an unknown of the equation. The trapezoidal family of methods, adapted for nonlinear problems, can be used for time stepping. In order to compute the displacement we use the finite difference formula of the Newmark family of methods.

2.4.1 Generalized trapezoidal family of methods

We wish to satisfy the following equation at time t_{n+1}:

$$\mathbf{M}(\mathbf{x}_{n+1})\mathbf{a}_{n+1} + \mathbf{K}(\mathbf{x}_{n+1})\mathbf{v}_{n+1} = \mathbf{F}_{n+1}^{ext} \tag{2.4.1}$$

The position $x_{n+1} = x_n + \Delta d_{n+1}$ is the (unknown) position at the end of step t_{n+1}.

The left-hand side can be grouped into a non-linear term N. This is done following the work by Hughes, Pister and Taylor [27]. The generalized trapezoidal algorithm can then be stated as:

$$N(a_{n+1}, v_{n+1}, x_{n+1}) = F^{ext}_{n+1} \tag{2.4.2}$$

with

$$a_{n+1} = \frac{1}{\Delta t \gamma}(v_{n+1} - \tilde{v}_{n+1}) \tag{2.4.3}$$

$$\Delta d_{n+1} = \Delta \tilde{d}_{n+1} + \Delta t^2 \beta a_{n+1} \tag{2.4.4}$$

and predictors given by

$$\tilde{v}_{n+1} = v_n + \Delta t(1-\gamma)a_n \tag{2.4.5}$$

$$\Delta \tilde{d}_{n+1} = \Delta t v_n + \frac{\Delta t^2}{2}(1-2\beta)a_n \tag{2.4.6}$$

Remark 2.4.1 *Note that this formulation is similar to the one used by Idelsohn et al. [31], before he introduces the time splitting. Choosing his parameter $\theta = 1$ and $\gamma = 1$ in our formulation the two are identical.*

Now we like to write the system of equations in terms of increments of velocities. Linearization of N at iteration i yields

$$\frac{\partial N(a^i_{n+1}, v^i_{n+1}, x^i_{n+1})}{\partial v^i_{n+1}} \Delta v = F^{ext}_{n+1} - N(a^i_{n+1}, v^i_{n+1}, x^i_{n+1}) \tag{2.4.7}$$

with

$$\frac{\partial N(a^i_{n+1}, v^i_{n+1}, x^i_{n+1})}{\partial v^i_{n+1}} = \frac{\partial(M(x^i_{n+1})a^i_{n+1})}{\partial v^i_{n+1}} + \frac{\partial(K(x^i_{n+1})v^i_{n+1})}{\partial v^i_{n+1}} \tag{2.4.8}$$

In this work we assume that the two matrices M and K don't change during an iteration, that is we assume that:

$$M(x^{i+1}_{n+1}) = M(x^i_{n+1}) \tag{2.4.9}$$

$$K(x^{i+1}_{n+1}) = K(x^i_{n+1}) \tag{2.4.10}$$

With this assumption the tangent matrix in Equation 2.4.8 becomes

$$\frac{\partial N(a^i_{n+1}, v^i_{n+1}, x^i_{n+1})}{\partial v^i_{n+1}} \approx M(x^i_{n+1})\frac{1}{\Delta t \gamma} + K(x^i_{n+1}) \tag{2.4.11}$$

2.4 – Time integration scheme

1. At t_{n+1} do:

   Initialize the iteration counter:
   $$i = 0$$

   Predictor phase:
   $$\mathbf{x}_{n+1}^{i=0} = \mathbf{x}_n + \Delta \tilde{\mathbf{d}}_{n+1}$$

   $\boxed{\text{Mesh update}}$

   $$\mathbf{v}_{n+1}^{i=0} = \tilde{\mathbf{v}}_{n+1}$$
   $$\mathbf{a}_{n+1}^{i=0} = 0$$

2. Compute the residual force, the tangent stiffness matrix and solve the linear system of equations:
   $$\Delta \mathbf{F} = \mathbf{F}_{n+1}^{ext} - \mathbf{N}(\mathbf{a}_{n+1}^i, \mathbf{v}_{n+1}^i, \mathbf{x}_{n+1}^i)$$
   $$\mathbf{K}^* = \frac{1}{\Delta t \gamma} \mathbf{M}(\mathbf{x}_{n+1}^i) + \mathbf{K}(\mathbf{x}_{n+1}^i)$$
   $$\mathbf{K}^* \Delta \mathbf{v} = \Delta \mathbf{F}$$

3. Corrector phase:
   $$\mathbf{v}_{n+1}^{i+1} = \mathbf{v}_{n+1}^i + \Delta \mathbf{v}$$
   $$\mathbf{a}_{n+1}^{i+1} = \frac{1}{\Delta t \gamma}(\mathbf{v}_{n+1}^{i+1} - \tilde{\mathbf{v}}_{n+1})$$
   $$\mathbf{x}_{n+1}^{i+1} = \mathbf{x}_{n+1}^i + \Delta \mathbf{d}_{n+1}^{i+1} = \mathbf{x}_{n+1}^i + \frac{\Delta t \beta}{\gamma} \Delta \mathbf{v}$$

   $\boxed{\text{Mesh update}}$

4. Test if computation has converged: If $|\Delta \mathbf{F}| < C \in \mathbb{R}$, go to 1. (step $n = n+1$). Else go to 2. (iteration $i = i+1$).

Table 2.1: Generalized trapezoidal algorithm.

Remark 2.4.2 *This approximation introduces a small error into the numerical method. It is the same approximation as the one followed by Idelsohn and coworkers in [31] for updated Lagrangian free-surface flows.*

The algorithm is summarized in Table 2.1.

Remark 2.4.3 *The algorithm presented in Table 2.1 has been chosen in view of being applied to the two-phase formulation presented in the following chapter. Since the two-phase formulation has to provide a framework for various types of material behavior, from viscous fluids to solidified material, the time stepping algorithm has to be able to accommodate a displacement-dependent stiffness term. With such an additional term the algorithm can easily be adapted to the Newmark family of methods, with parameters γ and β. We choose a backward Euler method, with $\gamma = 1$, and $\beta = 0.5$ for the computation of the displacements. Most numerical tests, unless mentioned otherwise, are performed using $\gamma = 0.9$ and $\beta = 0.49$. The results are almost identical (see solitary-wave test in Section 2.6.6).*

2.4.2 Mesh update

For the sake of generality we perform two updates at time step t_{n+1}: The first update at the beginning of the time step sets the spatial coordinates of the nodes to the predicted positions $\tilde{\mathbf{x}}_{n+1}$. The second update sets the spatial coordinates to the corrected positions $\mathbf{x}_{n+1}^{i+1}$ according to the new solution. In order to reduce computational cost the update for the predictor at the next step and the update after the last iteration at the previous step can be combined. After the nodes have been updated the domain has to be re-meshed and the variables need to be re-mapped onto the new nodes. The procedure is summarized in Table 2.2.

2.5 Spatial discretization and numerical approximation

The spatial discretization of the fluid domain consists of representing the continuous material by a set of discrete nodes, at which the solution of the governing equations are computed. The nodes, together with a set of interpolation functions to interpolate the solution between the nodes, define the numerical approximation to the boundary value problem.

We introduce the same numerical approximation to the velocity field $\mathbf{v}$ and the pressure field p

$$v_i^h(\mathbf{X}) = N^I(\mathbf{X}) v_i^I \quad (2.5.1)$$
$$p^h(\mathbf{X}) = N^I(\mathbf{X}) p^I \quad (2.5.2)$$

2.5 – Spatial discretization and numerical approximation

1. Update the spatial coordinates of the nodes

    (a) Predictor step:

    $$\tilde{\mathbf{x}}_{n+1} = \mathbf{x}_{n+1}^{i=0} = \mathbf{x}_n + \Delta \tilde{\mathbf{d}}_{n+1}$$
    $$= \mathbf{x}_n + \Delta t \mathbf{v}_n + \frac{\Delta t^2}{2}(1 - 2\beta)\mathbf{a}_n \qquad (1)$$

    (b) Corrector step:

    $$\mathbf{x}_{n+1}^{i+1} = \mathbf{x}_{n+1}^{i} + \Delta \mathbf{d}_{n+1}^{i+1} = \mathbf{x}_{n+1}^{i} + \frac{\Delta t \beta}{\gamma} \Delta \mathbf{v} \qquad (2)$$

2. Find the boundary of the fluid, using the α-shape method (see Appendix A.2)

3. Re-mesh inside the boundary (re-zone the nodes and create new triangles, see Section 2.5.4)

4. Re-map the nodal variables on the new mesh (see Section 2.5.5)

Table 2.2: Mesh update algorithm.

where v_i^I are the components of $\mathbf{v}^I$, the velocity and p^I the pressure at node I. $N^I(\mathbf{X})$ is the shape function of node I evaluated at the material coordinate $\mathbf{X}$. It is important to note that the shape functions are functions of material coordinates. When taking material derivatives with respect to time only the nodal values are affected while the shape functions remain constant.

$$a_i^h(\mathbf{X}) = N^I(\mathbf{X})a_i^I = N^I(\mathbf{X})\frac{Dv_i^I}{Dt} \qquad (2.5.3)$$

In the updated Lagrangian formulation spatial derivatives are taken with respect to the last known configuration $\Omega(X)$, which is identical to the current configuration $\Omega(x)$ at the end of the last time step. We can therefore write

$$\nabla \mathbf{v}^h(\mathbf{X}) = \frac{\partial N^I(\mathbf{X})}{\partial \mathbf{X}} \mathbf{v}^I \qquad (2.5.4)$$

A matrix form can be obtained by substituting the above trial functions together with the corresponding test functions into Equation 2.3.21. The global matrix equation (Equation 2.3.24) then becomes

$$\begin{bmatrix} \mathbf{M} & 0 \\ 0 & 0 \end{bmatrix} \begin{Bmatrix} \mathbf{a} \\ 0 \end{Bmatrix} + \begin{bmatrix} \mathbf{K} & \mathbf{G} \\ \mathbf{G}^T & 0 \end{bmatrix} \begin{Bmatrix} \mathbf{v} \\ p \end{Bmatrix} = \begin{Bmatrix} \mathbf{f} \\ 0 \end{Bmatrix} \qquad (2.5.5)$$

2.5.1 Numerical approximation

We seek an approximation that minimizes the distance between the field variables $\mathbf{v}$ and p and their approximations $\mathbf{v}^h$ and p^h throughout the domain. For fixed meshes finite element approximations are well adapted and have many desirable properties, such as convergence (the error decreases as the mesh is refined), compact support (shape functions are defined on an element, and are zero elsewhere), the interpolation property ($N^I = 1$ at node I and $N^I = 0$ at all other nodes) or the partition of unity property (in each point the sum of all shape functions is equal to unity).

2.5.1.1 Finite element method

The computational domain is decomposed into non-overlapping elements. Each element is defined by its nodal connectivity. On each element a local coordinate system is introduced, based on which the local approximation, the shape functions, are defined. These shape functions have local support, i.e. they are zero outside the element. The shape functions defined on an element are of polynomial form. For an isoparametric linear triangular finite element we define the following mapping between local coordinates ξ and η and global material coordinates X and Y

$$X = \xi X_1 + \eta X_2 + (1 - \xi - \eta) X_3 \qquad (2.5.6)$$
$$Y = \xi Y_1 + \eta Y_2 + (1 - \xi - \eta) Y_3 \qquad (2.5.7)$$

2.5 – Spatial discretization and numerical approximation

Figure 2.1 shows the mapping of the parent domain of a linear triangular finite element to the global domain. The coordinates (ξ, η) are also called barycentric coordinates. In the parent

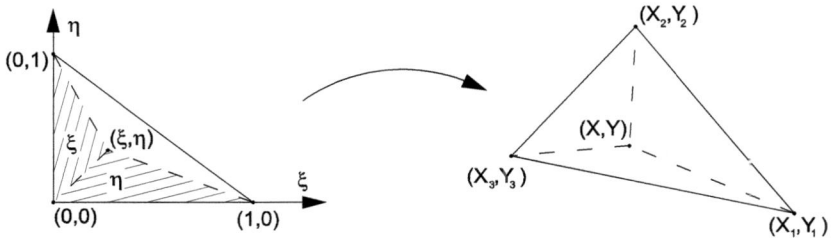

Figure 2.1: Mapping between parent and global domain for the linear isoparametric triangular finite element.

triangle they represent the areas of the sub-triangles as shown in the figure. The computation of barycentric coordinates for an arbitrary point (X, Y) is described in Table 2.3. For the linear

> The barycentric coordinates of a node (X, Y) with respect to a triangle Δ_{123} (see Figure 2.1) can be computed as follows:
>
> $$d_{11} = (X_3 - X_1)(X_3 - X_1) + (Y_3 - Y_1)(Y_3 - Y_1)$$
> $$d_{12} = (X_3 - X_1)(X_2 - X_1) - (Y_3 - Y_1)(Y_2 - Y_1)$$
> $$d_{13} = (X_3 - X_1)(X - X_1) + (Y_3 - Y_1)(Y - Y_1)$$
> $$d_{22} = (X_2 - X_1)(X_2 - X_1) - (Y_2 - Y_1)(Y_2 - Y_1)$$
> $$d_{23} = (X_2 - X_1)(X - X_1) + (Y_2 - Y_1)(Y - Y_1)$$
> $$D = d_{11}d_{22} - d_{12}^2$$
> $$\xi = \frac{d_{22}d_{13} - d_{12}d_{23}}{D}$$
> $$\eta = \frac{d_{11}d_{23} - d_{12}d_{13}}{D}$$

Table 2.3: Computation of barycentric coordinates.

triangle one Gauss point with the barycentric coordinates $\eta = \xi = 1/3$ is used. The shape functions and their derivatives for this element are given by

$$\mathbf{N} = \begin{bmatrix} \phi^1 \\ \phi^2 \\ \phi^3 \end{bmatrix} = \begin{bmatrix} \xi \\ \eta \\ 1 - \xi - \eta \end{bmatrix} \quad (2.5.8)$$

$$\nabla \mathbf{N} = \begin{bmatrix} \phi^1_{,\xi} & \phi^1_{,\eta} \\ \phi^2_{,\xi} & \phi^2_{,\eta} \\ \phi^3_{,\xi} & \phi^3_{,\eta} \end{bmatrix} = \begin{bmatrix} 1 & 0 \\ 0 & 1 \\ -1 & -1 \end{bmatrix} \quad (2.5.9)$$

In order to show the form of the shape functions we show the interpolated value on a rectangular grid, where all nodal values are equal to 0 except at one node, where the value is 1 (Figure 2.2). The derivatives in x-direction are given in Figure 2.3.

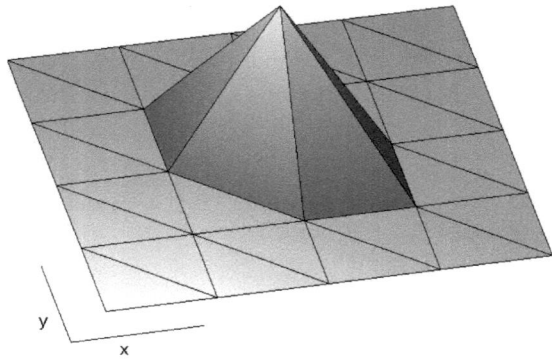

Figure 2.2: Finite element shape function.

A well-known problem of finite element modeling using Lagrangian formulations occurs if the nodes of the mesh undergo large motions. Individual elements can deform excessively until they eventually become invalid. Numerically this becomes the case if the determinant of the Jacobian $det(J)$, given in Equation 2.5.10, becomes negative.

$$\det(J) = \det \begin{bmatrix} \frac{\partial X}{\partial \xi} & \frac{\partial X}{\partial \eta} \\ \frac{\partial Y}{\partial \xi} & \frac{\partial Y}{\partial \eta} \end{bmatrix} = (X_1 - X_3)(Y_2 - Y_3) - (Y_1 - Y_3)(X_2 - X_3) \quad (2.5.10)$$

For such a situation it would be desirable to have a numerical approximation that is defined solely based on the positions of the nodes, not requiring any connectivity information. Such methods are generally referred to in the literature as "meshless methods".

2.5 – Spatial discretization and numerical approximation

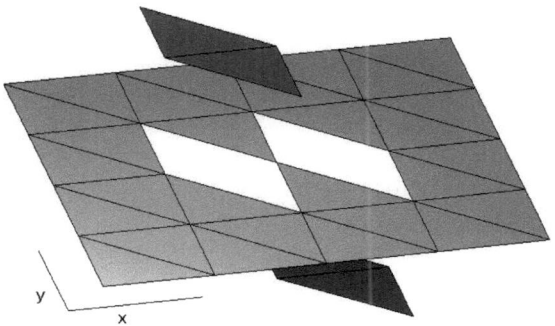

Figure 2.3: Finite element shape function derivative.

Meshless methods have been a subject of active research for many years. The idea in the first developments such as Smoothed Particle Hydrodynamics (SPH) was to formulate interaction between particles, without having to construct a mesh. This way all the information such as mass, velocity, temperature, is transported with the particle and no internal variables have to be stored at integration points. Particle methods are in general formulated in an updated Lagrangian framework where the equations are written in a frame of reference attached to the particles. While in the beginning meshless methods were mainly used to solve problems of a discrete nature, such as in astrophysics or granular flow, they have recently become increasingly popular in the context of continuum problems.

The main idea is to construct interpolations based on the distance to nodes in a neighborhood. Many different variants of meshless methods exist, for a review of early work see for example Belytschko et al. [3]. In our work we are interested in a method, that allows to approximate the solution of a boundary value problem in a discrete set of nodes, without having to be concerned about constructing a mesh. The most straightforward method to achieve this is point collocation, where a set of differential equations is solved without the need of having to integrate a weak form (see, for example Oñate et al. [41]). Problems with stability in collocation methods have, however, led us to explore meshless methods that make use of a weak form. In the following we present the Natural Element Method (NEM) which shares many desirable properties with the finite element method, such as compact support, the interpolation or Kronecker-delta property and the partition of unity property.

2.5.1.2 Natural Element Method

The Natural Element Method has been introduced by Sukumar et al. ([49] and [50]). It uses the geometrical concept of natural neighbors in order to define an interpolation that is solely based on point locations. The concept of natural neighbors emerges from the Voronoi diagram of a set of points. A Voronoi cell of a node I includes all points that are closer to node I than to any other node. Figure 2.4 illustrates a set of nodes with its corresponding Voronoi diagram (solid

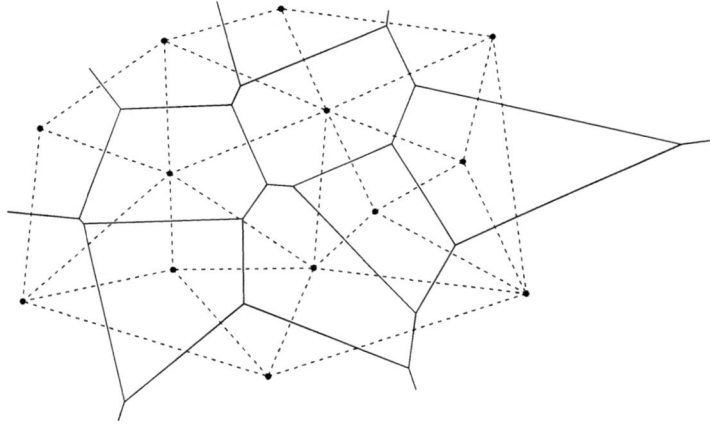

Figure 2.4: Voronoi diagram (solid lines) and Delaunay triangulation (dashed lines) for a set of nodes.

lines) and its underlying Delaunay triangulation [2] (dashed lines). Natural neighbors of a node are all nodes whose Voronoi cells are adjacent to the Voronoi cell of the node.

The construction of the Sibson interpolant for any point x is illustrated in Figure 2.5. It involves the construction of a secondary Voronoi diagram (dashed lines) and the computation of the areas of polygons resulting from intersecting the (secondary) Voronoi cell around point x with the (primary) Voronoi cells attributed to each natural neighbor of point x. The Sibson interpolant is then computed as

$$\Phi_i = \frac{A_i}{\sum_{k=1}^{nNN} A_k} \qquad (2.5.11)$$

---
[2] See Appendix A.2 for a description of Delaunay triangulations

2.5 – Spatial discretization and numerical approximation 23

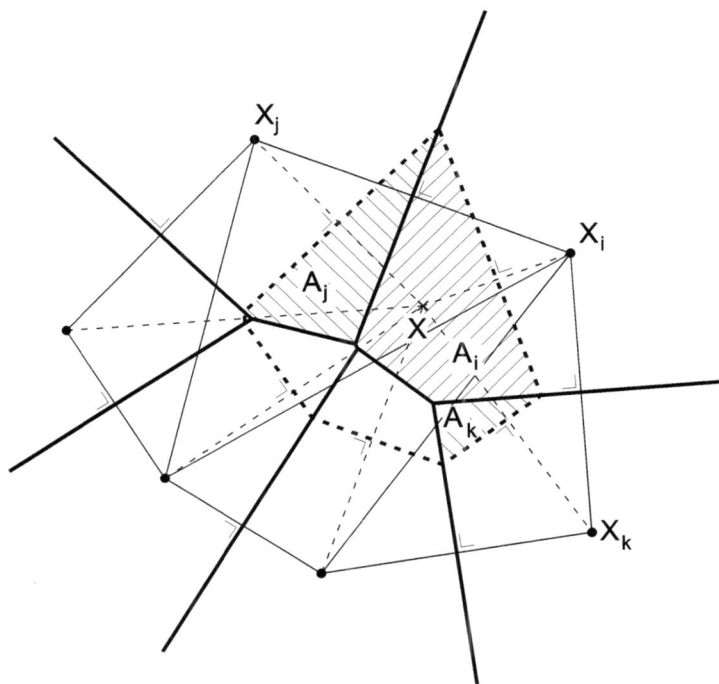

Figure 2.5: Computation of Sibson interpolant at a point **X**; Voronoi edges (thick lines) and Delaunay triangulation (thin lines). After insertion of a point **X** a secondary Delaunay triangulation and Voronoi diagram is constructed (dashed thick and thin lines). A_i, A_j and A_k stand for the areas of the hatched polygons and are used in the computation of the shape functions corresponding to nodes $\mathbf{X}_i$, $\mathbf{X}_j$ and $\mathbf{X}_k$.

where A_i are the areas as defined in Figure 2.6 and nNN is the number of natural neighbors of point x.

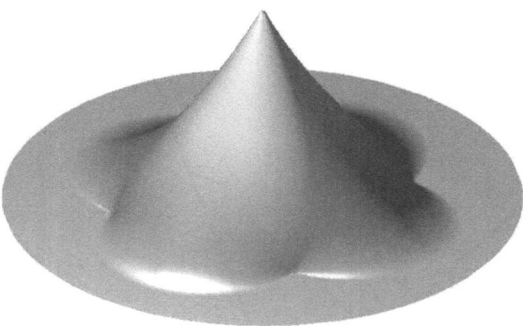

Figure 2.6: Sibson interpolant.

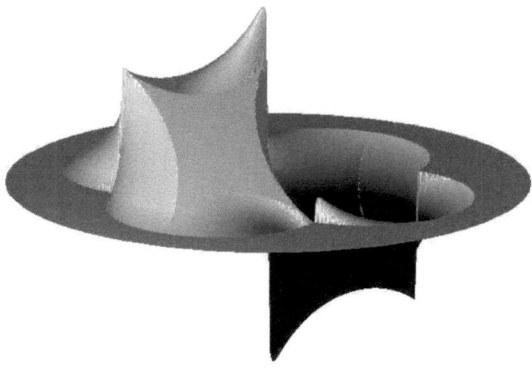

Figure 2.7: Derivative of Sibson interpolant.

The Sibson interpolant on a square grid interpolating a value of 1 at the center node and values of 0 at all other nodes is shown in Figure 2.6, the derivative in Figure 2.7. It can be seen that the shape function has a compact support that coincides with the union of circles circumscribed around the Delaunay triangles that have the central node as a vertex.

2.5 – Spatial discretization and numerical approximation

In order to assemble the global stiffness matrix we have to integrate local, or elemental, stiffness matrices on subdomains of a background mesh. In the case of NEM we use a Delaunay triangulation, since this structure already is available from the evaluation of the shape functions. On the triangles of the background mesh we evaluate the elemental matrices by Gauss quadrature. Other techniques exist, for instance nodal integration techniques (an overview of nodal integration techniques can be found in [44]) or Monte-Carlo methods. While Monte-Carlo methods suffer from high computational cost and low accuracy nodal integration techniques seem to be a viable alternative, if one needs to avoid the construction of a mesh by all means. The elemental matrices obtained from Gauss quadrature are larger than those obtained from triangular finite elements, because the computation of NEM shape functions depends not only on the nodes of the triangle, but on all the natural neighbors of the Gauss points of the triangle.

Meshless shape functions are in general rational functions, as can be seen in Equation 2.5.11. Gauss quadrature allows to exactly integrate polynomial functions. For rational functions, however, it can only approximate the integral. Even by using a very high order Gauss integration rule meshless shape functions can therefore not be integrated exactly. Additionally to the error introduced by approximating the exact solution v and p by v^h and q^h an error stemming from inexact evaluation of the integrals is introduced by the method. This can be illustrated as follows: We wish to compute the mass matrix M^e of an element e. Using Gauss quadrature on n_Q quadrature points, we have (see Belytschko, Liu and Moran [4], p. 166)

$$\mathbf{M}^e = \rho \int_{\Omega_e} \mathbf{N}^T \mathbf{N} d\Omega = \rho \sum_{Q=1}^{n_Q} w_Q \mathbf{N}^T(\mathbf{X}_Q) \mathbf{N}(\mathbf{X}_Q) d\Omega \qquad (2.5.12)$$

where $\mathbf{X}_Q$ are the coordinates and w_Q the weights of the quadrature points. This Equation is exact if the shape functions $\mathbf{N}$ are polynomials of order $m \leq 2 n_Q - 1$. For NEM shape functions Equation 2.5.12 is only satisfied approximately.

The result is that meshless methods cannot represent linear solution fields exactly. This is observed in a linear patch test, which will not be passed by most meshless methods.

2.5.2 Mesh-independent finite element method

The computation of NEM shape functions requires the construction of a Voronoi tessellation. Because the shapes of the Voronoi cells depend only on node locations the main requirement for a method to be called meshless is met. The construction of a Voronoi tessellation produces as a by-product a Delaunay triangulation, therefore it is a natural choice to use the triangulation as a background mesh for numerical integration. During the implementation of the NEM we decided to implement linear triangular finite element shape functions based on the Delaunay triangulation for comparison. The main difference between the two methods turned out to be the computation of the shape functions. All other steps in the solution procedure are identical. The claim is therefore that *triangular finite elements defined on a continuously updated Delaunay triangulation have the same meshless characteristics as the NEM*.

The use of meshless methods for updated Lagrangian simulations is often justified by the lower sensitivity to irregular meshes (e.g. [21]). The presence of interior angles to triangles of the mesh that are close to $180°$ deteriorates the accuracy of the results obtained by the finite element method. Such large angles are, however, a result of irregularly spaced nodes, as can be seen in Figure 2.8. In a) a regular triangular mesh is shown. The situation in b) illustrates a

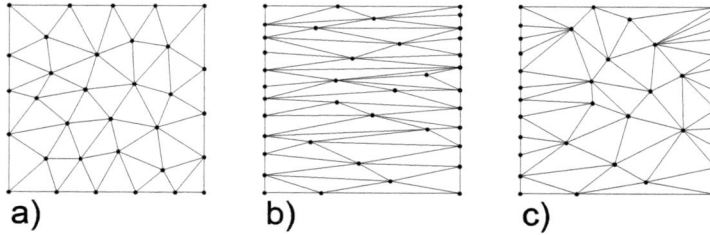

Figure 2.8: Regularity of meshes: a) Regularly spaced nodes, b) Irregularly spaced nodes, c) Irregularly spaced nodes with regularized triangles.

typical distribution of nodes after a horizontal stretching motion. Here some elements clearly are badly shaped as they have interior angles close to $180°$. In c) a new triangulation based on the same nodes as in b) is constructed. Far fewer triangles are badly shaped, in fact only triangles located on the boundary still have relatively large interior angles. Changing from mesh b) to mesh c) involves no re-mapping of nodal variables, as the nodal positions are not changed. In fact step b) to c) is exactly what has to be done when new NEM shape functions based on the nodal set in b) are to be computed. So the problem with badly shaped finite elements, that is triangles with interior angles close to $180°$, only occurs on the boundary.

Instead of using a meshless method we deal with the problem of uneven distribution of nodes by a re-meshing strategy. Then standard finite elements can be used that pass a linear patch test and have well established convergence properties. Moreover, as we will see in the next chapter, re-meshing is specifically required by the Lagrangian update of the nodes of the two-phase formulation and will therefore not create any additional computational overhead. For the single-phase formulation we claim that finite elements with re-meshing is more accurate than NEM.

2.5 – Spatial discretization and numerical approximation

The central question is: *What is the best way to obtain accurate solutions, if the nodes are distributed unevenly?* In our opinion the answer is to remediate the uneven distribution by remeshing. Then standard finite elements can be used that pass a linear patch test and whose convergence properties are well established. We claim that this method is more accurate for simulating free-surface fluid dynamics problems than NEM.

2.5.3 Volumetric locking and stabilization of the velocity-pressure mixed formulation

The existence of a unique finite element solution that converges at optimal rate depends on the satisfaction of the so called Ladyshenkaya-Babuška-Brezzi condition, a.k.a. the LBB-condition. The mathematical foundations of the condition, that sometimes is also called the *inf-sup condition*, can be found in Babuška [2] and Brezzi [6]. For a more application-oriented perspective of the LBB-condition and its implications for modeling of incompressible media we suggest reading of chapter 4 in Hughes [24]. The LBB condition postulates that certain combinations of velocity and pressure interpolation functions lead to volumetric locking in the velocity field and spurious oscillations in the pressure field. In particular this is the case for interpolations of equal order for the velocities and the pressure, which is most convenient from the point of view of implementation.

For a long time it was believed that meshless methods alleviate volumetric locking because the shape functions depend on more nodes than in commonly used finite elements. Further investigations showed that volumetric locking, and spurious pressure oscillations, can still occur if the approximation is built on a smaller domain of influence. This has been analyzed in Dolbow et al. [19] for EFG. For the NEM enriched shape functions that do not exhibit locking have been introduced by González et al. [22]. Special treatment of the incompressibility constraint is still required. Meshless methods, as compared to finite elements, therefore don't offer any new alleviated way to avoid the locking phenomenon.

Many common finite elements, such as the triangle with linear velocities and pressure, don't satisfy the LBB-condition and therefore suffer from volumetric locking. Other elements, for instance the linear triangle with nodes located at midsides, cannot be used for free surface flow as triangles on the free surface, which are connected to the rest of the domain by only one edge, would be allowed to rigidly rotate around the midnode of the shared edge (see Hughes [24] p. 250). An alternative triangular element that does not lead to spurious pressure oscillations at low velocities is the MINI-element, used in Braess et al. [5].

As an alternative to using interpolation functions that satisfy the LBB-condition the finite element solution to a boundary value problem can be stabilized by adding additional terms to the weak form or by solving for velocities and pressure separately. For the use in our model we investigated several stabilization methods, namely finite increment methods, pressure-stabilizing Petrov-Galerkin methods, Galerkin/least-squares methods and fractional step methods.

Finite increment methods have been developed by Oñate and co-workers. The basic idea is to take into account the discrete nature of the solution spaces. When computing derivatives of the governing equations over a finite element (or a finite increment) higher order terms of a Taylor expansion are retained which serve as stabilizing terms in the weak form. It is important to note that these higher order terms represent a measure of the deviation of the discrete solution from the exact solution. Adding such terms to the weak form can be seen as penalizing it with the local error, therefore forcing the discrete solution to approach the exact solution. As the mesh is refined the stabilization terms go to zero, an important property of any stabilization method. A detailed description of finite increment calculus for stabilizing incompressible solid and fluid mechanics problems can be found in Oñate et al. [42].

Pressure-stabilizing Petrov-Galerkin methods follow a similar rationale. Here terms resulting from multiplying the momentum equation with the gradient of the pressure test function are added to the weak form (Hughes et al. [25]).

$$B(\mathbf{w}^h, q^h; \mathbf{v}^h, p^h)_{PSPG} = B(\mathbf{w}^h, q^h; \mathbf{v}^h, p^h)$$
$$- \sum_e^{n_{el}} \tau_e \left(\int_{\Omega^e} \nabla q^h \cdot (\rho \frac{D\mathbf{v}^h}{Dt} - \nabla \cdot \tau(\mathbf{v}^h) - \nabla p^h) d\Omega \right) \quad (2.5.13)$$

$$L(\mathbf{w}^h, q^h)_{PSPG} = L(\mathbf{w}^h, q^h) - \sum_e^{n_{el}} \tau_e \left(\int_{\Omega^e} \nabla q^h \cdot \mathbf{b} d\Omega \right) \quad (2.5.14)$$

τ_e is a stabilization parameter for element e defined further down in the text. Such a method satisfies consistency, because the stabilization terms disappear if $\mathbf{v}^h$ and q^h are replaced by $\mathbf{v}$ and q. The additional terms penalize strong pressure gradients by the residual of the momentum equation and spurious oscillations of pressures are therefore avoided. Because the test functions are different from the trial functions it is considered a Petrov-Galerkin method.

Remark 2.5.1 *When linear approximations are used the term $\nabla \cdot \tau(\mathbf{v}^h)$ cancels because it involves second derivatives. In meshless methods this term has to be computed for the method to be consistent.*

Remark 2.5.2 *If we omit the inertial term (as in the steady state case) and the body force term in Equation 2.5.14 then we recover the original stabilization method by Brezzi and Pitkäranta [7].*

Galerkin/least-squares methods form a general framework that includes pressure-stabilizing Petrov-Galerkin and puts them into a context with other stabilization methods such as streamline-upwind Petrov-Galerkin, if an Eulerian reference frame is used. Stabilization is achieved by

2.5 – Spatial discretization and numerical approximation

adding least-squares terms of the residual of the momentum equation to the weak form (see Hughes, Franca and Hulbert [28]).

$$B(\mathbf{w}^h, q^h; \mathbf{v}^h, p^h)_{GLS} = B(\mathbf{w}^h, q^h; \mathbf{v}^h, p^h)$$
$$+ \sum_e^{n_{el}} \tau_e \left(\int_{\Omega^e} (\rho \frac{D\mathbf{w}^h}{Dt} - \nabla \cdot \tau(\mathbf{w}^h) - \nabla q^h) \cdot (\rho \frac{D\mathbf{v}^h}{Dt} - \nabla \cdot \tau(\mathbf{v}^h) - \nabla p^h) d\Omega \right) \quad (2.5.15)$$

$$L(\mathbf{w}^h, q^h)_{GLS} = L(\mathbf{w}^h, q^h)$$
$$+ \sum_e^{n_{el}} \tau_e \left(\int_{\Omega^e} (\rho \frac{D\mathbf{w}^h}{Dt} - \nabla \cdot \tau(\mathbf{w}^h) - \nabla q^h) \cdot \mathbf{b} d\Omega \right) \quad (2.5.16)$$

Remark 2.5.3 *In the steady-state case and when linear finite elements are used, the Galerkin/least-squares method is identical to pressure-stabilizing Petrov-Galerkin, with the exception of the sign of the stabilization term.*

Fractional step methods take another approach to stabilizing the pressure for transient incompressible flow. By separating the momentum equation and the continuity equation and solving separately for velocities and pressures satisfaction of the LBB-condition can be avoided, at least as long as the time step doesn't become too small. An in-depth discussion of the LBB-condition in the context of fractional step methods can be found in Guermond et al. [23].

Many different variations of fractional step methods exist, the interested reader can for instance consult Codina [12] and references therein for a more thorough analysis. Idelsohn et al. [31], who we referenced earlier, use the scheme proposed by Codina. The main idea is to split the momentum equation into two parts, use the first part to compute an intermediate velocity while using the previous pressure, and then use the continuity equation to compute the new pressure. Finally the second part of the split momentum equation is used to compute the new velocity at the end of the time step. A by-product of this splitting of the system of equations is the reduction of the number of equations that have to be solved simultaneously. However, additional approximations introduced by these methods leads to some loss of accuracy.

For the present problem we seek a numerical method that gives stable solutions while being easy to integrate in an updated Lagrangian setting. We expect the pressure-stabilizing Petrov-Galerkin method to best meet these requirements.

2.5.3.1 Stabilization parameter for pressure-stabilizing Petrov-Galerkin

After having settled for a specific method the stabilization parameter τ_e used in Equations 2.5.13 and 2.5.14 has to be defined. For incompressible flow in the steady-state regime the parameter from [25] can be used:

$$\tau_e = \frac{\alpha h_e^2}{2\mu} \qquad (2.5.17)$$

where α is a dimensionless parameter and h_e a characteristic element diameter. For unstructured meshes generated with a Delaunay mesher h_e is set equal to the maximum triangular side length. This definition, however, doesn't account for the inertial forces present in the transient regime. We use the modified stabilization parameter given in Tezduyar et al. [54]:

$$\tau_e = \frac{1}{\sqrt{\left(\frac{2\rho}{\Delta t}\right)^2 + \left(\frac{2\mu}{\alpha h_e^2}\right)^2}} \qquad (2.5.18)$$

Δt is the time step length.

With these stabilization terms the weak form is now complete. The global matrix equation 2.5.5 is modified as follows:

$$\begin{bmatrix} M & 0 \\ 0 & 0 \end{bmatrix} \begin{Bmatrix} a \\ 0 \end{Bmatrix} + \begin{bmatrix} K & G \\ G^T & S \end{bmatrix} \begin{Bmatrix} v \\ p \end{Bmatrix} = \begin{Bmatrix} f \\ f_s \end{Bmatrix} \qquad (2.5.19)$$

where S and f_s are the stabilization terms arising from Equations 2.5.13 and 2.5.14. Details on computing the individual sub-matrices in Equation 2.5.19 are given in Appendix A.1.

2.5.4 Re-meshing

Re-meshing of the computational domain assures a mesh of good quality throughout the simulation. Due to the update of the nodal coordinates the finite elements deform and will eventually become invalid, unless the domain is re-meshed. Re-meshing consists of two steps: Re-zoning of the nodes, and re-creating the elemental connectivities. Both steps don't have to be performed after each Lagrangian update. At relatively small mesh deformation all elements remain well-shaped, i.e. all element Jacobians are greater than zero. Neither re-meshing nor re-zoning has to be performed. As deformation increases, some elements become badly shaped (Jacobian $J \leq 0$) and new element connectivities based on the same set of nodes have to be established. Re-zoning is required only at very large deformation, if nodes become clustered along lines due to stretching in one direction. Re-zoning the nodes in order to create an evenly spaced nodal set restores a good quality approximation.

For the sake of generality we re-mesh in this work after every iteration. More sophisticated methods can be obtained by using indicators of mesh distortion or error estimators. Methods of selective local re-meshing can be found for example in Malcevic et al. [35] or Radovitzky et al. [45].

2.5 – Spatial discretization and numerical approximation

2.5.5 Re-mapping procedure

After a new, evenly spaced set of nodes has been created all the nodal variables have to be mapped onto the new nodes. We accomplish this task by linear interpolation, thus using the same interpolation as in the finite element approximation.

Proposition 2.5.4 *The solution obtained from the finite element method using linear shape functions converges quadratically with respect to the mesh size. Linear interpolation of the solution onto a new set of nodes also converges at least quadratically. The re-mapping therefore doesn't introduce an error of order lower than 2 and quadratic convergence of the numerical method is retained. This is shown numerically in Section 2.6.3.*

Remapping of masses is not required in this case since the fluid is considered incompressible and the mass density therefore constant. The nodal masses for the next iteration can be obtained by simply computing lumped mass matrices on the new nodal configuration.

The velocities and accelerations, the pressure and the incremental solution vector $[\Delta \mathbf{v} \; \Delta p]^T$ have to be mapped from the previous mesh to the new mesh at the current iteration. The re-mapping procedure consists of:

For each node (X, Y) of the new mesh, do:

- Find the triangle Δ_{123} of the previous mesh, that contains node (X, Y).
- Compute the barycentric coordinates ξ and η of node (X, Y) with respect to triangle Δ_{123} as explained in Table 2.3. The barycentric coordinates correspond to the weights that are used to interpolate a variable at node (X, Y) within the nodes of triangle Δ_{123}.
- Interpolate all variables ϕ at node (X, Y): $\phi = \xi \phi_1 + \eta \phi_2 + (1 - \xi - \eta) \phi_3$

In order to test if a point (X, Y) is contained in a triangle with vertices (X_1, Y_1), (X_2, Y_2) and (X_3, Y_3) we can again use the barycentric coordinates: If $min(\xi, \eta, 1 - \xi - \eta) > 0$, then the point is located inside the triangle. In order to narrow down the number of triangles to test to those that are close to point (X, Y) we use the method for spatial search described in Section 4.3.

2.5.6 Detection of contact at the fluid boundary

Nodes on the free surface can eventually come in contact with a fixed boundary as the flow mass propagates along its downhill path. We chose to deal with this contact problem in a very simple way. Figure 2.9 illustrates the algorithm for the nodal update, contact detection at the boundary and correction of the coordinates of the nodes in contact with the boundary. The fixed boundary is specified as a sequence of straight line segments. As the Lagrangian update

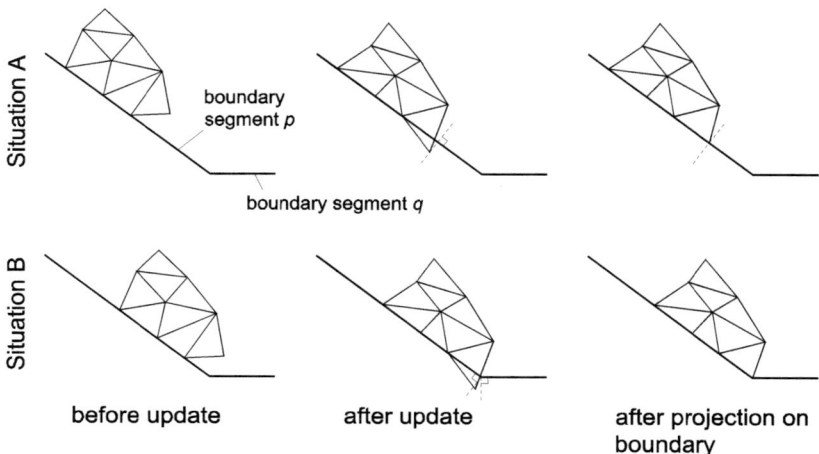

Figure 2.9: Algorithm for detection of contact with boundary and correction, illustrated for the situation of contact along a straight line and in a corner.

of the nodal positions is carried out we test for all nodes whether their new location is above or below the fixed boundary. If it is below, then the node is projected perpendicularly onto the nearest fixed boundary segment (Situation A). If the node is below a segment, but in a corner (Situation B), the node is projected back to the corner. This relatively crude method results in a small loss of mass every time a node is projected back onto the fixed boundary. Also it produces shocks and small pressure jumps. However these effects appear to be acceptable in the context of our simulations.

2.6 Numerical tests

The single-phase model is verified and validated according to the program of tests shown in Table 2.4.

2.6.1 Hydrostatic patch test

A hydrostatic patch test is performed on an inclined, fluid-filled container (Figure 2.10). The fluid under a body load of $\mathbf{b} = [0\ -10]^T$ is at rest, or in other words, the velocity is equal to zero

2.6 – Numerical tests

Single-phase tests	Aspect to be verified	Comparison with
Hydrostatic patch test	Interpolation	Exact result
Transient lid-driven cavity flow	Stabilization	Steady state norm (Equation 2.6.3)
Stratified flow	Mass flux at inflow vs. outflow, re-mapping on fixed mesh, convergence	Exact result
Flow over step	Performance at high Re, rezoning on fixed mesh	Denham et al. [16]
Flow over step with free surface	Free surface	Qualitatively
Solitary wave	Conservation of energy	Energy balances (Equation 2.6.9)
Droplet formation	Surface tension	Qualitatively
Dam break	Rezoning, surface tension	Mass conservation, CPU-time

Table 2.4: Verification tests for single-phase model.

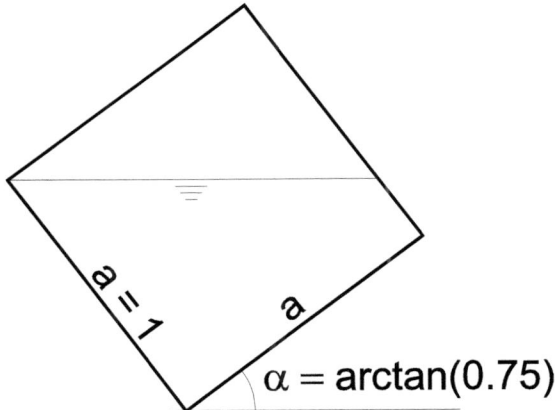

Figure 2.10: Geometry used in the hydrostatic patch test.

while the pressure distribution is linear (hydrostatic). The density is $\rho = 1000$ and the viscosity $\mu = 0.001$. Even though this is a static problem, it is analyzed using the transient algorithm described earlier. One time step of $\Delta t = 0.001$ is computed. The test is performed using linear triangular finite elements and the NEM with two different Gauss quadrature rules. The L_2 error norms for the pressure and velocity, given in Table 2.5, are defined for scalars and vectors as:

$$||q||_{L_2} = \sqrt{\int_\Omega q^2 \, d\Omega} \qquad (2.6.1)$$

$$||\mathbf{q}||_{L_2} = \sqrt{\int_\Omega (\mathbf{q} \cdot \mathbf{q}) \, d\Omega} \qquad (2.6.2)$$

	h	$||p_{exact} - p_h||_{L_2}$	$||\mathbf{v}_{exact} - \mathbf{v}_h||_{L_2}$
Finite elements	0.1	$6.8 \cdot 10^{-7}$	$6.6 \cdot 10^{-18}$
NEM 1 GP	0.1	$6.9 \cdot 10^{-1}$	$4.8 \cdot 10^{-4}$
	0.05	1.7	$8.3 \cdot 10^{-4}$
	0.02	8.8	$2.6 \cdot 10^{-3}$
NEM 25 GP	0.1	$1.2 \cdot 10^{-1}$	$2.0 \cdot 10^{-4}$
	0.05	$4.6 \cdot 10^{-1}$	$4.1 \cdot 10^{-4}$
	0.02	3.1	$1.2 \cdot 10^{-3}$

Table 2.5: Error norms for the hydrostatic patch test. Comparison between FEM and NEM, using Gauss quadrature with 1 and 25 integration points.

While the linear finite elements pass this patch test the natural neighbor-based method doesn't. Not only is the latter unable to represent exactly a linear pressure field, the error even increases as the mesh is refined. Clearly the velocity field that results from a NEM computation shows spurious, meaningless oscillations, arising from the fact that the weak form cannot be integrated exactly even if the exact solution is linear. Using a higher order integration rule decreases the error only marginally. While not-fulfillment of a linear patch test doesn't necessarily cause problems in the case of a solid mechanics application it is clearly unacceptable for an updated Lagrangian method. These spurious velocities will be used to update the nodal positions at the end of every time step and the error will therefore accumulate with time.

2.6.2 Transient lid-driven cavity flow

The stabilization added to the method modifies the equation of conservation of mass, introducing a small compressibility to the formulation. One effect of this artificial compressibility

2.6 – Numerical tests

is a delay in how fast a steady state flow regime can be reached. This is tested on a lid-driven cavity flow test, which is a well studied problem for incompressible elasticity as well as incompressible fluid dynamics. The test consists of a fluid filled squared domain on which a constant horizontal velocity is imposed on the top boundary. On all other boundaries zero velocity is imposed. Figure 2.11 shows the mesh used in all computations. Since the boundary of the

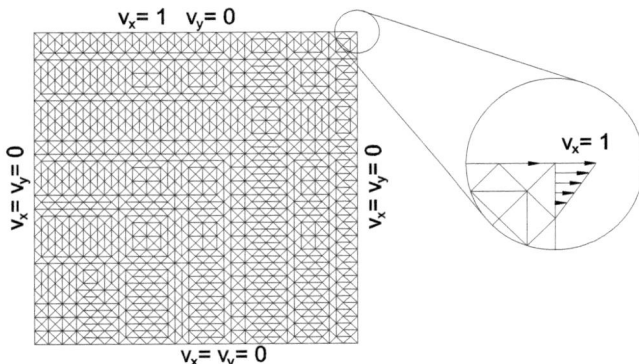

Figure 2.11: Mesh and boundary conditions used in the computations for the cavity-flow problem. The mesh is composed of 1105 nodes.

computational domain stays fixed throughout the analysis we use the same fixed finite element mesh, on which the nodal variables are mapped back after each Lagrangian update. Starting from a fluid at rest the convergence toward the steady state solution is examined by looking at the change in velocity between two consecutive time steps:

$$e_{steady} = \frac{||\mathbf{v}_h^{n+1} - \mathbf{v}_h^n||_{L2,\Omega}}{||\mathbf{v}_h^{n+1}||_{L2,\Omega}} \qquad (2.6.3)$$

Three sets of analyses are performed: One where the complete momentum equation is included in the pressure gradient stabilization (denoted c in the results), and two where the inertial term is dropped. These two analyses are performed for two different mass densities of the fluid: $\rho_1 = 1$ and $\rho_2 = 100$. The viscosity μ is in all analyses equal to unity. The following stabilization parameters have been tested: $\alpha = [0.001, 0.05, 1., 20]$. The computations are performed on a mesh consisting of 1105 nodes, using a time step of length $\Delta t = 0.05$. The results in Figure 2.12 show the steady state error on a logarithmic scale plotted as a function of time.

Remark 2.6.1 *For all methods the solution tends toward a steady state. The rate at which the solution converges is higher for problems at low Reynolds numbers, where inertial effects are less important.*

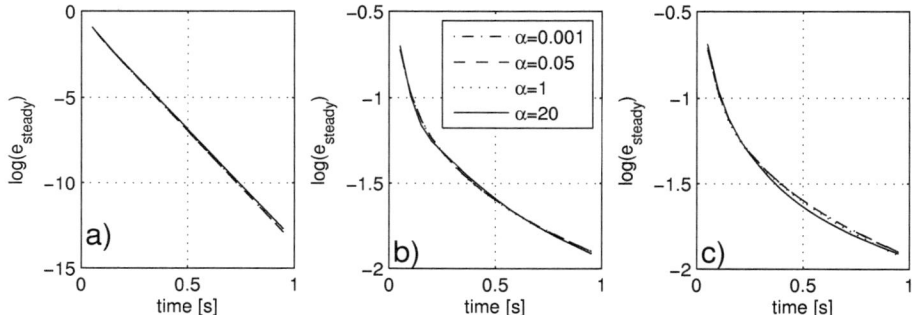

Figure 2.12: Steady state convergence for cavity flow: a) $\rho = 1$, b) $\rho = 100$, c) consistent stabilization with $\rho = 100$.

Remark 2.6.2 *The consistent stabilization, that is when the inertial term in Equation 2.5.13 is included in the stabilization, does not improve the rate at which the solution converges to a steady state.*

Remark 2.6.3 *The value of the stabilization parameter appears to have very little influence on this particular problem. In the following we use $\alpha = 0.5$.*

2.6.3 Stratified flow

With this relatively simple test we want to show that the solution converges with optimal rate. The test is performed on the geometry illustrated in Figure 2.13. Again, the nodal variables

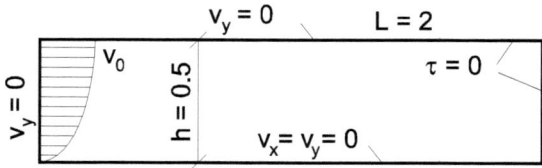

Figure 2.13: Geometry used in the stratified-flow test.

are mapped back on the same fixed mesh. At the inflow section a parabolic distribution of horizontal velocity with a maximum velocity of $\mathbf{v}_x = 1$ and $\mathbf{v}_{x,y} = 0$ at the top is imposed. The vertical velocity component along all four boundaries is set equal to zero. On the lower horizontal boundary the horizontal velocity is set to zero while at the upper boundary the

2.6 – Numerical tests

tangential force is zero. At the outflow boundary zero pressure is imposed. The analysis is performed for Reynolds numbers of 10 and 1000. The definition of the Reynolds number is given by

$$Re = \frac{\bar{v}\rho D}{\mu} \qquad (2.6.4)$$

where $\bar{v}$ is the cross-sectional average of the velocity, ρ the density, μ the dynamic viscosity and D a characteristic length of the cross section, in this case the height of the fluid domain. The density is $\rho = 10$ in all analyses, while the viscosity values of $\mu_1 = 0.01$ and $\mu_2 = 1$ are taken to vary the Reynolds number. The results for four structured meshes with 226, 502, 851 and 1814 nodes are computed, using a time step length of $\Delta t = 0.02$.

Remark 2.6.4 *Such a test with well defined boundaries fixed in space would typically be performed in an Eulerian description. Using an updated Lagrangian description we need to pay special attention to the inflow and outflow boundary conditions. A material point located on the inflow boundary at a time t_{n+1} was outside the computational domain at time t_n. Imposing a constant velocity in time on the inflow boundary therefore implies that the velocity is constant along streamlines in upstream direction. In the current problem this can be assumed to be true as the velocities in the exact solution are purely horizontal and identical for all vertical sections. For more complex flows we need to consider perturbed zones near the inflow and outflow boundaries where neglecting to convect velocities introduces a small error.*

In Figure 2.14 we show the evolution with time of the L_2 error norm of the velocity for Reynolds numbers of 10 and 1000. Due to the compressibility introduced by the stabilization

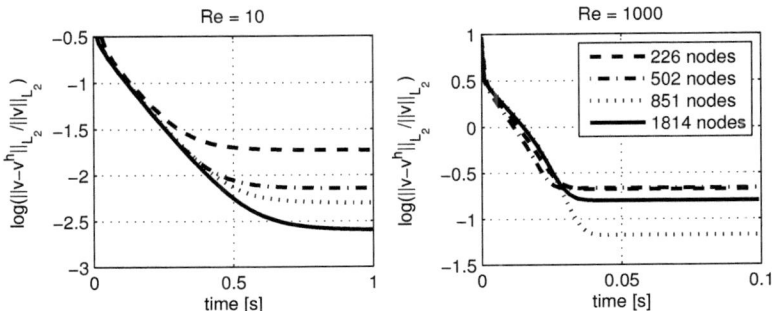

Figure 2.14: Evolution of the velocity error for stratified flow.

steady state is not reached immediately. After a certain time the error reaches a plateau, which can be considered steady state. For a given discretization the steady state error cannot get smaller than this value. The residual errors at steady state are plotted against the numbers of nodes in a logarithmic plot in Figure 2.15. For $Re = 10$ the convergence rate is quadratic.

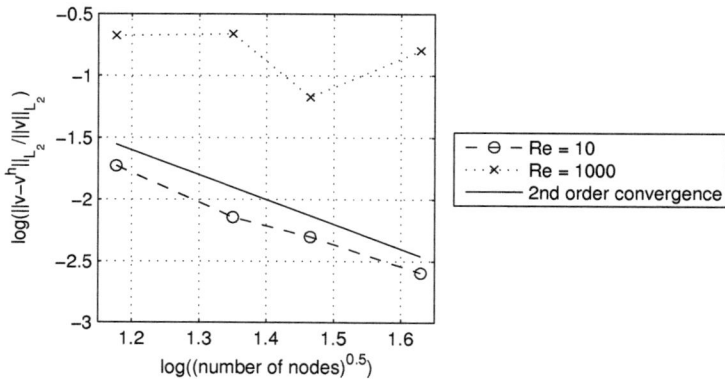

Figure 2.15: Convergence of the velocity error for stratified flow.

However, at $Re = 1000$ the solution doesn't converge at all. If we look at a contour plot of the velocities after steady state has been reached in the $Re = 1000$ test (see Figure 2.16) we can see that the vertical velocity is non-zero over a large portion of the domain. This is a result of the compressibility introduced by the stabilization. Since this error propagates from time step to time step we cannot expect to conserve optimal rates of convergence throughout the analysis. In this perspective the rate at which the error of the $Re = 10$ test tends to zero can be considered a good result.

In problems that involve the motion of a free surface mass conservation is essential. In order to obtain a measure for how well the mass in conserved we compute the difference between the volume of fluid entering and leaving the computational domain. For an incompressible fluid this difference should be equal to zero. We integrate the horizontal velocity along the in- and outflow boundaries:

$$q_{in} = \int_{\Gamma_{inflow}} v_x \, dS \qquad (2.6.5)$$

$$q_{out} = \int_{\Gamma_{outflow}} v_x \, dS \qquad (2.6.6)$$

$$error_{mass} = \frac{q_{out} - q_{in}}{q_{in}} \qquad (2.6.7)$$

2.6 – Numerical tests

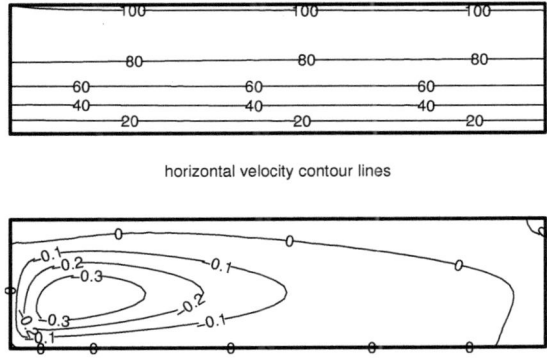

Figure 2.16: Contour lines of the horizontal and vertical velocities for stratified flow at $Re = 1000$.

In Figure 2.17 we show the mass error as a function of time for Reynolds numbers of 10, 100 and 1000. The time to reach steady state conditions decreases as the mesh is refined. At the same time the error in steady state decreases. Here we observe close to second order convergence for all Reynolds numbers, as can be seen in Figure 2.18.

2.6.4 Flow over a backward-facing step

In this simulation we test the mesh independent updated Lagrangian finite element method for incompressible flows at Reynolds numbers of 191 and 3015. For comparison we use experimental results obtained by Denham et al. [16] for turbulent flow and Denham et al. [17] for laminar flow.

The geometry of the problem consists of a narrow inlet part and a sudden enlargement after a backward-facing step (Figure 2.19). The boundary conditions are $v_x = 0$ and $v_y = 0$ everywhere, except at the outflow boundary where zero normal traction is specified and at the inflow boundary where the horizontal velocity profile is imposed. At the outflow boundary the pressure is set to zero. The Reynolds numbers in both computations are computed using the height of the step h_1 as characteristic length D ($Re = \bar{v}\rho D/\mu = 3015$). For both tests the results are shown after 1000 time steps. Table 2.6 summarizes the parameters.

For the test at $Re = 3015$ the inflow profile measured in [16] with an average velocity of $\bar{v} = 0.003015$, is prescribed. In Figure 2.20 the numerical results are compared to measured

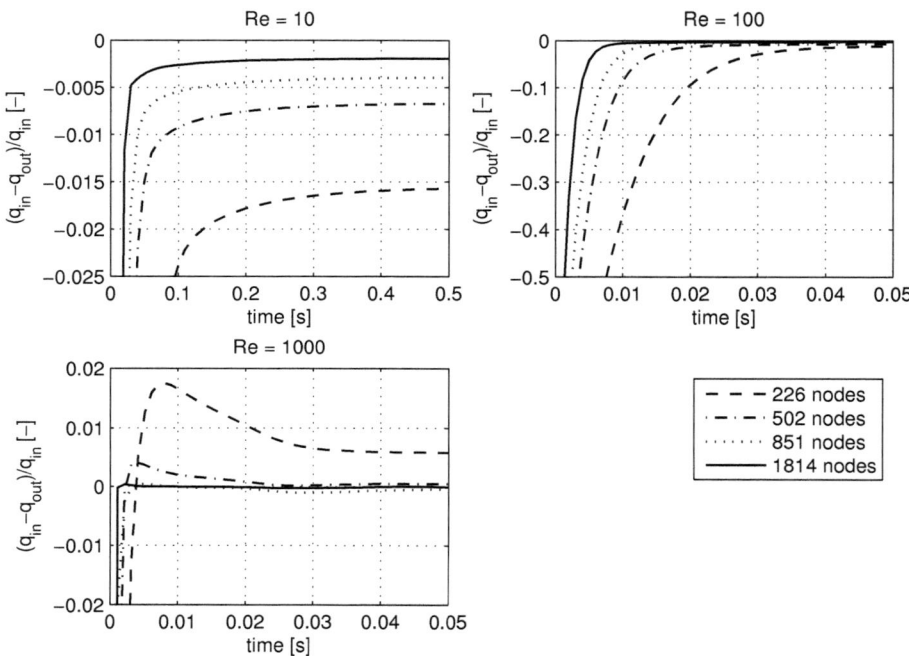

Figure 2.17: Mass error for stratified flow.

Re	$\overline{v}$	μ	ρ	Δt	N_{nodes}
191	0.287	0.01	10	0.2	7887
3015	0.00387	0.001	1000	50	2530

Table 2.6: Parameters used in backward-facing step problem.

2.6 – Numerical tests

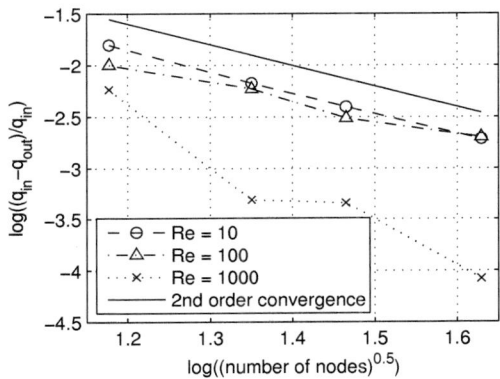

Figure 2.18: Convergence of mass error for stratified flow.

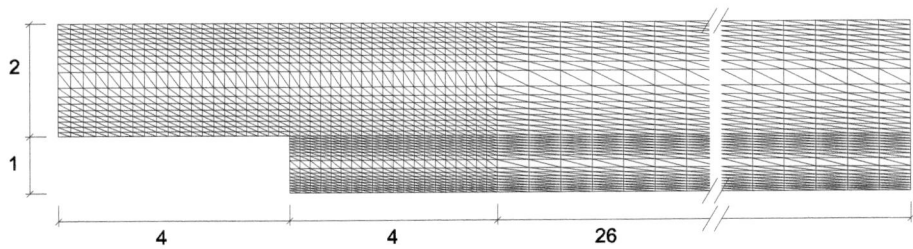

Figure 2.19: Mesh used in the backward-facing step problem at $Re = 3015$.

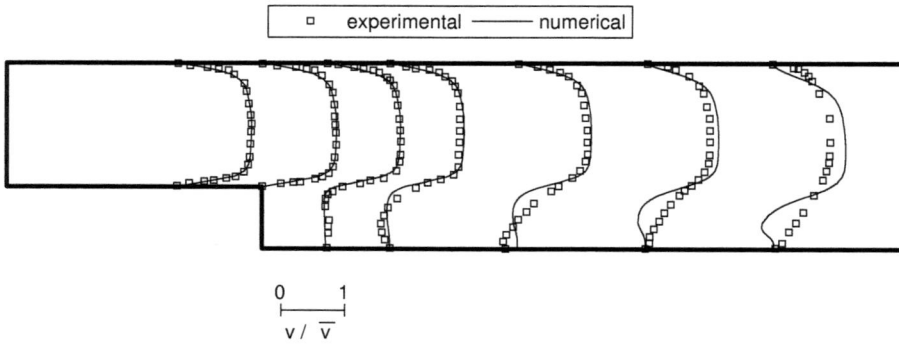

Figure 2.20: Comparison between experimental and numerical results for flow over a backward- facing step at $Re = 3015$.

values at several vertical sections. The agreement in terms of horizontal velocity is not very good, in particular the zone behind the step where a vortex is created extends too far to the right in the simulation. This discrepancy is a result of the inability of the simulation to capture the fine scales of the turbulent flow. Either a finer discretization near the boundaries, ideally using mesh adaptivity, could capture boundary layers more accurately, or a turbulence model could be adopted that would include the influence of the fine scales. For flows at high Reynolds numbers a method that is able to resolve the small scales of velocity fluctuations is indispensable.

Better results can be expected in the test where $Re = 191$. Here the velocity profile at the inflow section was set to a parabolic distribution with a maximum of 0.287. The result in Figure 2.21 shows a very good agreement between experimental and numerical results. This confirms that the numerical method performs well in the low Reynolds number range.

2.6.5 Flow over a backward-facing step with free surface

The next step in the verification and validation procedure is to test the updated Lagrangian formulation for representing an evolving free surface. The goal is to be able to obtain a smooth free surface without spurious oscillations or saw-tooth patterns. To this end we simulate flow over a backward-facing step, similar to the test in the previous section, but this time without the top boundary being fixed. The geometry shown in Figure 2.22 with the dimensions $L_1 = 0.5$, $L_2 = 1.5$ and $h_1 = h_2 = 0.25$) is used. At the inflow boundary a constant horizontal velocity

2.6 – Numerical tests

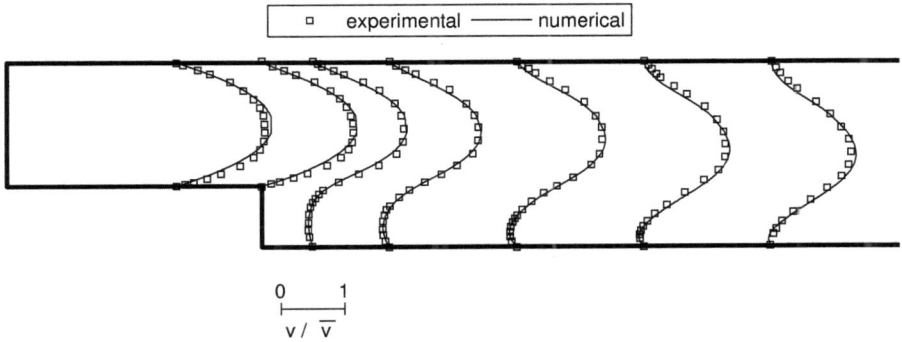

Figure 2.21: Comparison between experimental and numerical results for flow over a backward-facing step at $Re = 191$.

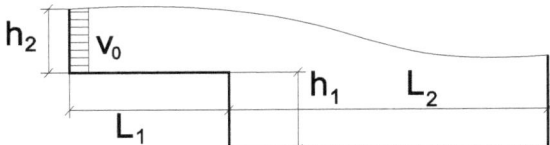

Figure 2.22: Geometry of the backward-facing step with free surface.

of $v = 2$ and a vertical velocity equal to zero are imposed. The zero vertical inflow velocity has as a consequence that the fluid depth at the inflow boundary is constant. At the outflow boundary the traction is set equal to zero and no Dirichlet condition is imposed on the pressure. The Reynolds number for this problem is $Re = 500$, using h_1 as the characteristic length of the problem. Figure 2.23 shows streamlines after 1000 time steps at $t = 2$ seconds when the

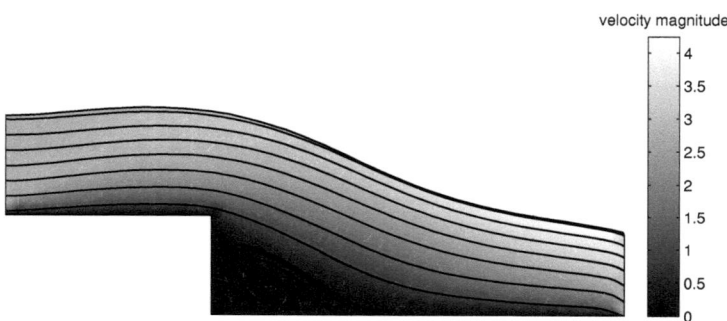

Figure 2.23: Streamlines and velocity magnitude for flow over a backward-facing step with free surface.

flow is approximately steady. The velocities are smooth and no oscillations are visible at the free surface. Behind the step a small-amplitude recirculation region appears. Right after the inflow boundary on the left side the initially constant velocity profile gradually adapts to a near-parabolic distribution. During this process the mean velocity decreases, which in turn raises the free surface. The overall qualitative aspect of this result doesn't reveal any deficiency in the numerical method.

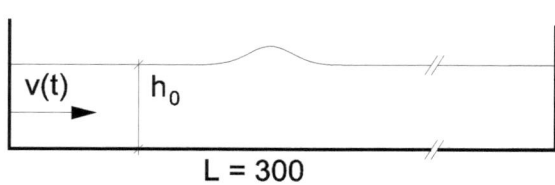

Figure 2.24: Geometry of the solitary wave problem

2.6 – Numerical tests

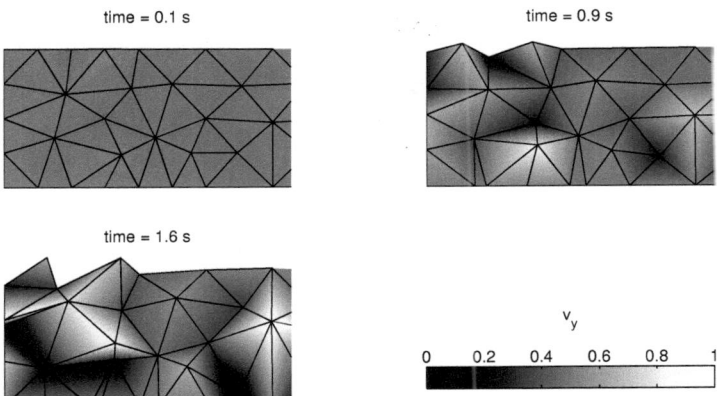

Figure 2.25: Solitary wave, using NEM: Zoom to the left boundary where the horizontal velocity is prescribed. Colors represent the vertical velocity field.

2.6.6 Solitary wave

In this test a horizontal displacement is prescribed to the side of a rectangular, fluid-filled container of a length of 300 (see Figure 2.24). This displacement creates a solitary wave propagating on the free surface from left to right, rebounds from the wall and travels back in the opposite direction. The displacement that we have to impose in order to create this very specific type of wave is taken from [43]:

$$d_x = \sqrt{\frac{4}{3}h_0\eta_0}\left(1+\tanh\left(\sqrt{\frac{3\,\eta_0}{4\,h_0^3}}ct-4\right)\right) \qquad (2.6.8)$$

where $\eta_0 = 0.86$ is the wave height, $h_0 = 10$ the fluid depth, t the time, $c = \sqrt{gh_0(1+\frac{\eta_0}{h_0})}$ the speed of propagation of the wave on the free surface and $g = 10$ the (vertical) gravitational acceleration. The analyses are performed using a time step $\Delta t = 0.1$ and a mesh composed of 337 nodes. The density ρ is chosen to be $\rho = 1$ and two values of viscosity, $\mu = 0.001$ and $\mu = 100$, are selected.

In this simulation we look at energy conservation for analyses using different values of viscosity. Before showing the results using finite elements we show the results of an attempt to employ the NEM to simulate this problem at $\mu = 0.001$. A zoom on the zone near the boundary where the velocity is imposed (Figure 2.25) shows strongly oscillating vertical velocity fields for three selected time steps. The analysis eventually has to be terminated due to the velocity field producing a very uneven fluid surface. Using a higher order integration rule doesn't improve the result. The problem is caused by the method not satisfying the patch test exactly (see Section 2.6.1) and precludes the use of NEM shape functions in similar types of low viscosity problems.

In the following we investigate conservation of energy more closely. To this end the internal, kinetic and potential energies are computed at each time step. The internal energy is the energy dissipated by viscous forces. The rate of change in total energy can be written as

$$\underbrace{\dot{E}}_{\text{total energy rate}} = \underbrace{\sigma : \nabla \mathbf{v}}_{\text{internal energy rate}} + \underbrace{\rho \mathbf{v} \cdot \mathbf{a}}_{\text{kinetic energy rate}} + \underbrace{\rho \mathbf{g} \cdot \mathbf{v}}_{\text{potential energy rate}} \qquad (2.6.9)$$

In order to obtain the change of total energy the kinetic energy rate is integrated symbolically ($\int \rho \mathbf{v} \cdot \mathbf{a} dt = 0.5 \rho \mathbf{v} \cdot \mathbf{v}$) whereas the internal and potential energy rates are integrated numerically. Figure 2.27 shows the partial and the total energies for the two analyses with different viscosities. We can see that the partial energies increase as mechanical work is put into the system by moving the left wall. The applied work is transformed into kinetic energy, potential energy, and a part of is is dissipated by viscous forces. As the wave reaches the wall on the right side after about 40 seconds the water level at the wall raises. This is caused by kinetic energy being transformed into potential energy. At about 44 seconds the water depth at the wall reaches its maximum while kinetic energy drops to zero (numerical value 0.42). The wave rebounds and continues to travel back and forth until all kinetic energy is eventually dissipated. In both analyses the free surface is perfectly smooth, which is a very good result in particular for the low-viscosity case. The free surfaces are shown in a surelevated view in Figure 2.26.

For the analysis at $\mu = 100$ the energy that is input into the system is almost immediately dissipated (transformed into internal energy or heat). Because the fluid offers more resistance to the applied movement of the wall if the viscosity is higher the total energy is slightly higher in the case of $\mu = 100$.

The exact potential energy after all kinetic energy has dissipated can be computed analytically. The total horizontal displacement by which the left wall is moved results in a raise in water depth and thus an increase of potential energy of 3462. This increase can be compared to the change in potential energy obtained from the computation with a high value of viscosity. The relative error of the potential energy computed numerically is only about 10^{-5}, after the kinetic energy has been dissipated almost completely at $t = 200$ seconds. The total energy, that is the sum of kinetic, potential and through viscous forces dissipated energy should remain constant throughout the analyses. This is approximately true for the high-viscosity analysis, whereas for the low-viscosity analysis we note a small drop of total energy. This drop can be attributed to numerical dissipation of the time integration algorithm.

2.6 – Numerical tests

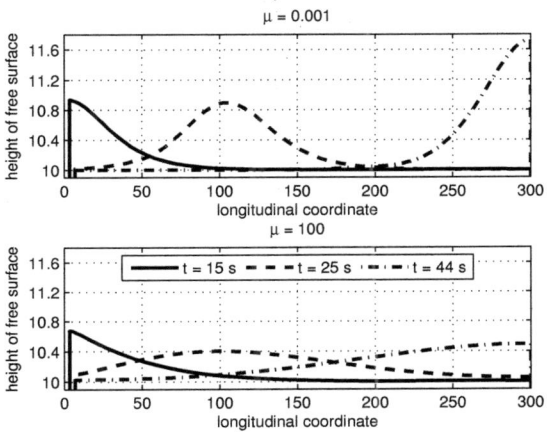

Figure 2.26: Energy balance for solitary wave problem.

In order to investigate numerical dissipation, we compare results for $\mu = 0.001$ obtained with the time stepping algorithm using various sets of parameters. The sets ($\gamma = 0.9$, $\beta = 0.49$), ($\gamma = 1$, $\beta = 0.5$) and ($\gamma = 0.5$, $\beta = 0.5$) are compared using two different time steps: $\Delta t = 0.1$ and $\Delta t = 0.2$. Figure 2.28 shows an excerpt of the total energy. The algorithm with $\gamma = 1$ dissipates slightly more energy than $\gamma = 0.9$. With $\gamma = 0.5$ the total energy is almost constant. The small bumps even decrease as the time step size is reduced. Neither of the algorithms show any oscillatory behavior. For $\gamma \neq 0.5$ numerical dissipation increases for larger time step sizes. In the following numerical tests the set of parameters ($\gamma = 0.9$, $\beta = 0.49$) is used. This somehow arbitrary choice is justified since for the use in general test cases, some dissipation of high frequencies is desirable.

An analysis with $\mu = 0.001$ and $\Delta t = 0.1$ but without re-zoning of nodes and re-mapping of variables yielded results that were indistinguishable from those shown in Figure 2.27.

2.6.7 Formation of drop due to surface tension effect

Free surface flows of viscous materials generally tend to preserve sharp angles at the free surface. This is the case for example in the dam-break problem as can be seen in Section 2.6.8. The reason for this anomaly is the fact that in a small element at the free surface the gravity load produces no shear forces that could deform the corner element sufficiently to smooth the cor-

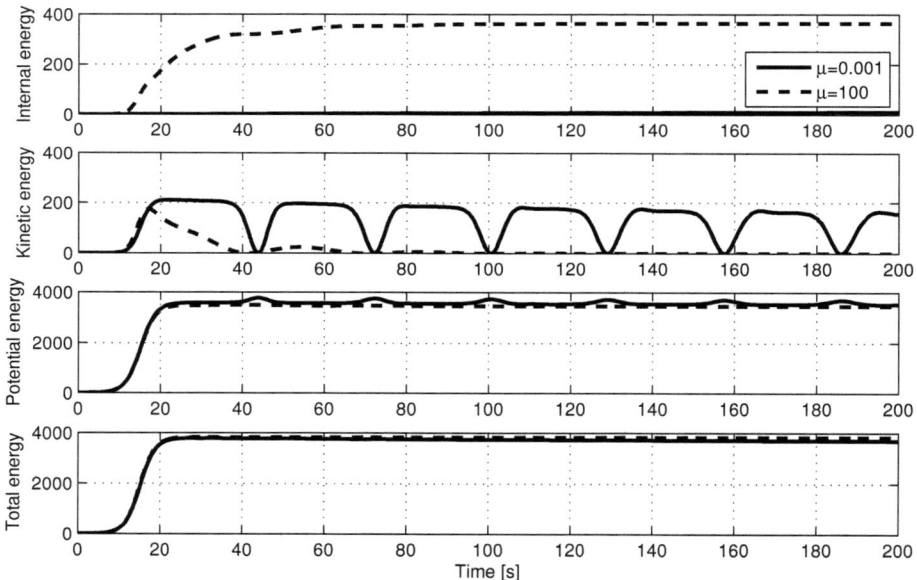

Figure 2.27: Energy balance for solitary wave problem.

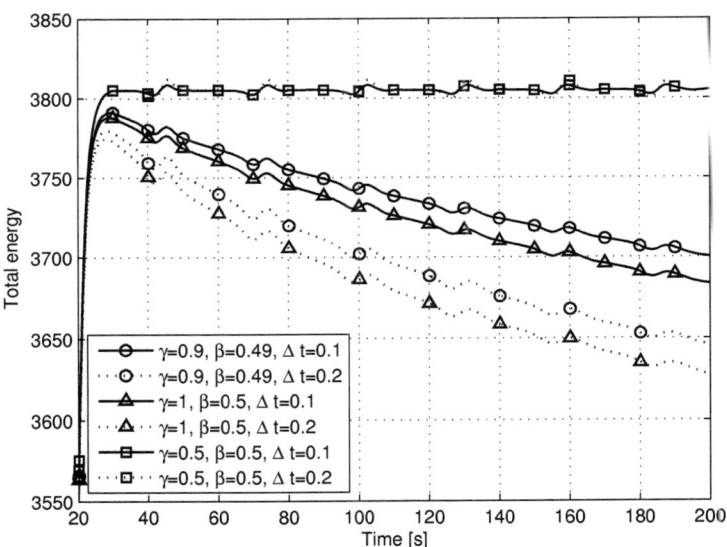

Figure 2.28: Detail of the total energy balance for the solitary wave problem.

ner out. The only way to obtain smooth edges in such a case is to include surface tension to the numerical model. A simple way to do this is by applying nodal forces normal to the boundary nodes that are proportional to the curvature of the boundary. The implementation follows the description of Caboussat [8].

$$\mathbf{F}_{fs} = \chi \kappa \mathbf{n} \tag{2.6.10}$$

where χ is a constant of proportionality, depending on the two media that are in contact at the

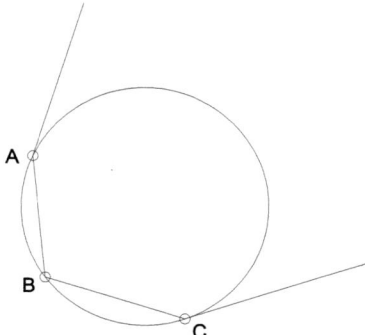

Figure 2.29: Computation of curvature in two dimensions.

boundary and has the unit of $\frac{N}{m}$. κ is the curvature ($\frac{1}{m}$) and n is the outward unit normal vector to the boundary. The curvature κ at a node can be estimated by computing the inverse radius of the circle passing through the node and its two neighbors on the boundary (see Figure 2.29). κ can then be computed as follows:

$$p = 0.5(d_{AB} + d_{BC} + d_{AC}) \tag{2.6.11}$$

$$\kappa = \frac{4\sqrt{(p(p-d_{AB})*(p-d_{BC})*(p-d_{AC})}}{d_{AB}d_{BC}d_{AC}} \tag{2.6.12}$$

d_{AB} is the distance between points A and B. For concave boundaries the curvature has to be multiplied by -1.

As a simple test for the implementation of the surface tension the formation of a droplet under gravity loading is computed. Figure 2.30 shows the deformation of the initially rectangular fluid into a drop. Initially sharp corners are completely smoothed out. After 20 seconds the drop is in equilibrium and the pressure distribution becomes hydrostatic.

2.6 – Numerical tests

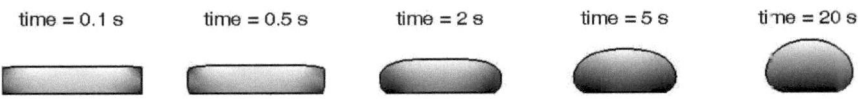

Figure 2.30: Droplet formation due to surface tension. Shading shows pressure distribution.

2.6.8 Dam break

A square block of fluid at rest is released by instantaneously removing the 'dam' on the right side (see Figure 2.31). This problem is characterized by the material undergoing large deforma-

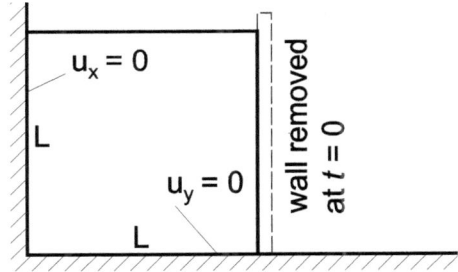

Figure 2.31: Geometry of the dam-break problem.

tions. We compare three strategies to deal with this problem: Finite elements with and without re-zoning, and NEM where the nodes are not re-zoned. In this study the density is set to $\rho = 1$ and the viscosity to $\mu = 100$. The time step length is chosen to be $\Delta t = 0.5$. The models consist of approximately 430 nodes.

The different approaches are compared in three ways. Figure 2.32 shows the relative error in conserving the volume of the fluid. While the results using FEM are almost identical, with a maximum relative error being smaller than $2 \cdot 10^{-5}$, the volume error using the NEM is considerably larger. Table 2.7 summarizes the volume errors as well as the runout distances (the maximum horizontal coordinate). Figures 2.33 and 2.34 show the meshes, colored by the pressure, at $t = 10$. It is interesting to see that the volume error for NEM doesn't decrease monotonically as we increase the order of the integration rule. Looking at Figure 2.34 it seems that the 3 Gauss-point rule yields less stiff results than the one- and 25-Gauss-point rules. To this point we have no explanation for this anomaly. It confirms nevertheless our choice of using

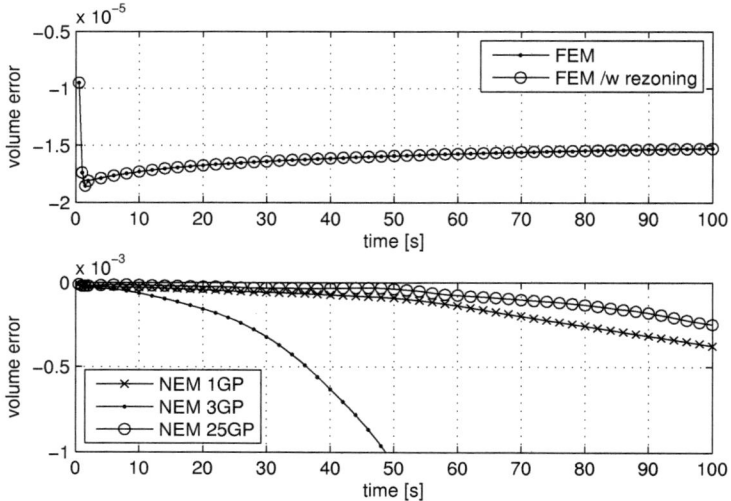

Figure 2.32: Dam break: Relative volume error.

Method	maximum relative volume error	runout distance
FEM without re-zoning	$-1.9 \cdot 10^{-5}$	2.545
FEM with re-zoning	$-1.9 \cdot 10^{-5}$	2.546
NEM 1 GP	$-3.8 \cdot 10^{-4}$	2.544
NEM 3 GP	$-3.9 \cdot 10^{-3}$	2.635
NEM 25 GP	$-2.5 \cdot 10^{-4}$	2.548

Table 2.7: Summarized results at $t = 10$ for the dam break test.

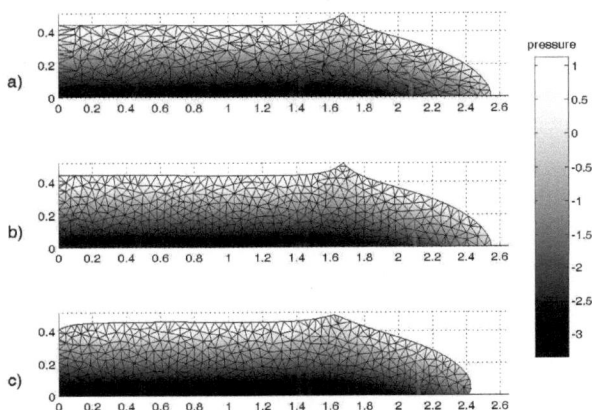

Figure 2.33: Dam break: Geometries at $t = 10$ seconds. a) FEM, b) FEM with re-zoning of nodes, c) FEM with surface tension.

finite elements rather than the natural neighbor-based method.

An additional result (Figure 2.33 c) points out the smoothing effect surface tension has on the free surface. The results a) to c) can not be distinguished by the bare eye, except for the distribution of nodes. Finally CPU-times are compared in Figure 2.35 for a series of analyses on meshes with varying numbers of nodes. The slope of the log-log plot for all three curves are approximately identical. In absolute values the FEM with re-zoning is the most expensive method, followed by the NEM, and FEM without re-zoning being the cheapest.

As a conclusion from these comparisons we can say that FEM with re-zoning is likely to be the most efficient while remaining accurate in the presence of highly distorted meshes, for which regular FEM would fail. The analysis using the NEM with one-Gauss-point integration has slightly faster CPU-time than FEM with re-zoning, but the much lower accuracy in terms of mass conservation is an important disadvantage for the analysis of free-surface flows.

2.7 Conclusions

It has been the goal of this chapter to establish the fundamentals of free-surface flow of an incompressible single-phase fluid. A mixed velocity-pressure formulation with stabilization

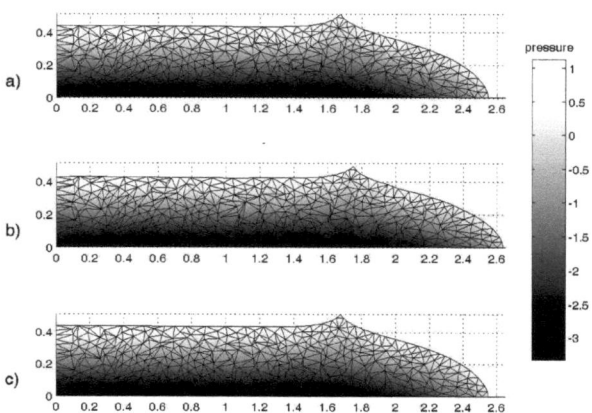

Figure 2.34: Dam break: Geometries at $t = 10$ seconds. Results obtained by using NEM: a) 1 Gauss point, b) 3 Gauss points, c) 25 Gauss points.

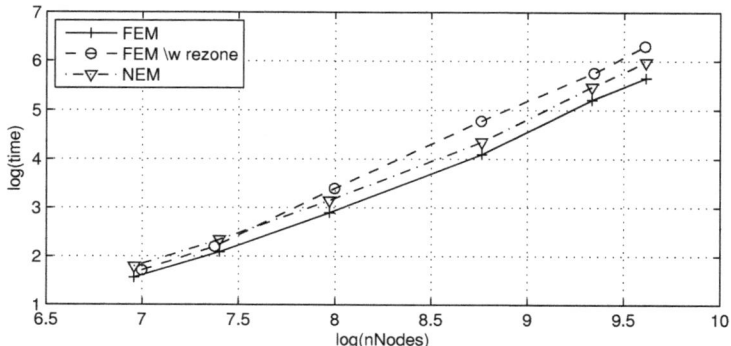

Figure 2.35: Scaling of CPU time for different discretization schemes.

2.7 – Conclusions

of the pressure has been proposed and implemented using linear triangular finite elements. The stabilization method has shown to suppress spurious pressure oscillations effectively in all numerical tests. The lid-driven cavity-flow test has allowed to select appropriate stabilization parameters and has shown that it is not mandatory for the stabilization to be consistent. All together the formulation is not very sensitive to the stabilization method, especially in the range of low Reynold's numbers.

The choice of an updated Lagrangian formulation has shown to be capable of accurately capturing the free surface of fluids. Smooth surfaces are obtained, even in the solitary-wave problem, where small perturbations could, due to the very low viscosity of the fluid, easily propagate and pollute the solution.

The updated Lagrangian formulation, in combination with re-meshing and re-mapping of nodal variables, yields second order convergence in terms of velocity up to the range of $Re = 10$ and in terms of the conserved mass up to the range of $Re = 100$. This very important result has been obtained on the example of stratified flow. Up to $Re = 200$ the method accurately reproduces experimental results, as has been shown in the test of flow over a backward-facing step. The type of mudflows that we are interested in modeling is within this range.

Triangular finite elements as well as a meshless method, the Natural Element Method (NEM) have been tested. Finite elements perform better than the NEM, in particular in terms of mass convergence. Instabilities at the free surface, as evidenced in the solitary-wave problem, further confirm the choice of the numerical approximation. While the ideas of meshless methods have had a considerable influence on the development of the approach, it has become apparent in the process that the main features of most meshless methods can be obtained using finite elements in combination with a re-meshing strategy.

Chapter 3

Updated Lagrangian method for modeling of multi-phase free-surface flows

3.1 Introduction

In this section we develop a two-phase model for simulating free-surface flows. The model has to be capable of tracking two different materials, constituents of a mixture, during their propagation down a slope. At each time step the model has to provide velocities of both phases, pressures and volume fractions everywhere in the mixture. The goal is to obtain a model providing a framework for the implementation of various types of material behavior. The target problem to be solved is the flow of two-phase mixture down a slope and its impact on an obstacle. From this analysis we want to obtain detailed time histories of forces acting on the obstacle.

3.1.1 Classification of mud- and debris-flow models

In a review paper from 1996, Hutter, Svendsen and Rickenmann summarize previous work in the field of numerical modeling of mud or debris flows and classify the various methods [30]. The authors chose a classification according to the structure of the underlying mathematical equations as well as according to the numerical procedure for spatial discretization.

Mathematical structure: A hierarchical structure can be established as follows: The simplest approach, a single-component model, employs the mass and momentum balance equations of a solid-fluid mixture. On the other side the two-component approach makes use of the individual mass and momentum balance equations of a solid and a fluid phase. It considers interaction between the phases through the solid and the fluid stress tensors as well as through a direct interaction force, also called drag force. A third category of approaches can be situated in between the two previous ones. Instead of using the balance equations of both phases only the mass balance of the solid phase and the mass and momentum balance equations of the mixture are taken into account.

Generally speaking the two-component approach can be custom tailored to take into account the largest variety of phenomena, such as erosion and sedimentation, while a single-component approach can effectively model situations where fluctuations of phase concentrations are not too important.

Spatial discretization: From simple mass point models to fully fledged three-dimensional debris-flow models several degrees of sophistication in terms of spatial discretization techniques can be found in the literature. Simple hydraulic models reduce the initially three-dimensional problem to two or one spatial dimensions. Averaging of the fluid-flow equations over the depth yields a family of methods that can accurately predict mean flow velocities, flow depths and run-out distances. Channel models are essentially one-dimensional in the direction of the flow path. These averaged models require special attention when complex geometries, surface effects such as base erosion or friction, or important variations of velocities, concentrations of solid particles or other quantities play a role. Finally three-dimensional models are capable to take into account the largest range of phenomena on complex terrain.

3.1.2 Review of mud- and debris-flow literature

Current state of the art in mud- or debris-flow modeling is almost exclusively based on depth-averaged models. The governing equations are integrated over depth and the depth of the flow, the velocity and the stress at the base are computed for each grid point. While some authors assume a hydrostatic pressure distribution (O'Brian et al. [40]) others assume a velocity profile and compute the pressure as a function of the averaged velocity (Savage et al. [46])

Single-phase models have been used extensively in the past for modeling of mudflows. Newtonian models that are used in hydraulics can be extended to account for the transport of solid material up to a certain volume fraction. These models can be refined by allowing the viscosity of the fluid to vary as a function of the solid volume fraction or the shear strain rate. Such constitutive models are referred to as non-Newtonian fluid models. Bingham fluids have a yield shear stress below which no flow occurs. They can be used for instance for simulating the stopping of a flow as the slope flattens out. A lot of work in this field has been done by Chen, see for instance [11, 10]. Many depth-averaged mud-flow models use a combination of Newtonian or non-Newtonian fluids together with a Mohr-Coulomb friction law at the base of the flow ([46, 40]). In current engineering practice the flood-routing software FLO-2D [1], which implements a depth-averaged formulation and offers a variety of constitutive models, is commonly used to predict flow-paths and establish hazard maps.

Single-phase models are however limited when it comes to predict complex behavior of mudflows without being able to calibrate the model. As Hutter points out in [29] this is mainly due to neglecting the effect of the interstitial fluid. The effect of pore pressure on the fluidization of a solid-fluid mixture is often indispensable for accurately describing the mechanics of mudflows.

Two-phase models have been around for a long time in the literature describing processes in nuclear engineering, where the behavior of air-water mixtures are analyzed. The efforts of the petroleum industry have also led to considerable progress in the field of modeling of multiphase fluids. The books by Soo [48] and Kolev [33] are two references worth citing. In the field of mud or debris flows Takahashi [53] was one of the first researchers recognizing the importance of taking into account the interaction between two phases and integrating it in a mathematical model. Svendsen et al. [52] laid out a thorough foundation for a general mixture

---
[1] www.flo-2d.com

of multiple phases by deriving all the balance equations from thermodynamical principles. Finally, Iverson and Denlinger [32, 18] generalized the depth-averaged model by Savage et al. [46] for the flow of a two-phase mixture over three-dimensional terrain.

In the process of depth averaging information on the distribution of stresses and velocities is lost. In order to compute forces on protection structures a detailed vertical profile of stress is required. Such detailed information can however only be obtained from full three-dimensional continuum models. Furthermore a continuum approach is the only sound way to obtain a model that can simulate all aspects of mudflows: Initiation of the flow, propagation, entrainment of material on the flow path, extent of spreading of the deposition zone, velocity of propagation and forces. Only a few such models, that offer the possibility for such capabilities to be integrated, are available in the literature. We note the work of Shao et al. [47] and Laigle et al. [34], who adopted a meshless method to simulate the impact of a mudflow on a structure. Both authors used a single-phase, non-Newtonian model.

3.1.3 Proposed model

In this chapter we develop a continuum theory for modeling the flow of a two-phase material. The scope of the model is for the purpose of testing limited to two spatial dimensions, although all the components are intended to be extensible to three dimensions. The mixture is assumed to be incompressible.

In the flow of a mixture of two phases the interaction between the phases becomes very important. We provide an algorithmic framework that tracks the evolution of two phases by their volume fractions. The starting point for our model is a single-phase viscous fluid. By choosing both phases to be viscous fluids we are able to match single-phase behavior as a limit case. A momentum exchange term models interaction between the two phases. Such a model should be capable of modeling the sedimentation of a denser phase within a less dense phase. This conceptually simple model allows to analyze its behavior on simple test problems, and can accommodate more complex material behavior without major changes to the formulation.

Similar to the single-phase fluid, the motion of the two-phase fluid is described in an updated Lagrangian frame of reference. The benefits for capturing the advancing front of a fluid have been clearly pointed out. Only the domain occupied by the flow is discretized. Since in an updated Lagrangian formulation the reference frame is attached to the material, and the material undergoes a different motion depending on what phase is considered, the update of nodal coordinates becomes more delicate than in the single-phase formulation. In our model we compute the velocities of both phases in each node of the mesh. The Lagrangian update of the coordinates of each node then leads to two different spatial coordinates for each phase.

3.1 – Introduction

Each Lagrangian update therefore doubles the number of nodes. At the next step we choose to create an entirely new mesh with about the same number of nodes as before the update. On this new mesh the phase velocities are mapped from the updated nodal coordinates. By doing so we are also able to conserve a mesh of good quality with even nodal spacing throughout the domain.

The physical description of the two-phase fluid is inspired by the definitions given by Kolev in [33] and Soo [48]. The phases are defined as follows:

- The fluid phase is a homogeneous mixture of water and microscopic solid particles. These solid particles are surrounded by water only.

- The solid phase consists of soil particles that are sufficiently large so they start to sediment during the time span considered for the analysis. The solid phase includes water contained in pores and a fine layer surrounding the soil particles.

We assume both phases to be present in the entire domain. This means that no phase interfaces need to be tracked, the two phases are smeared over the volume. Their corresponding presence is given by a volume fraction that takes values between 0 and 1. The volume fractions, which appear in all equations, are assumed to be constant during one iteration. The volume fractions are updated at the end of each iteration.

In order to remain most general in the description of the two-phase fluid a mathematical structure is chosen that uses the momentum balance equations of both the solid and the fluid phase. The mass balance equations of both phases are combined into one equation that enforces the conservation of the mass of the mixture. With the pressure being the same in both phases we have 5 equations for 5 nodal unknowns.

The modeling assumptions and their justifications are summarized in Table 3.1.

The mathematical description of the two-phase fluid model is given in the following. The presence of phase $p \in \{s, f\}$, where indices s and f stand for the solid and the fluid phase, is given by the volume fraction C_p:

$$C_p = \frac{V_p}{V} \qquad (3.1.1)$$

where V is the control volume and V_p is the volume within V that is occupied by phase p. From assumption D it can be concluded that

$$C_s + C_f = 1 \qquad (3.1.2)$$

Considering a control volume that is fixed in space another important observation can be made:

$$\frac{\partial C_s}{\partial t} + \frac{\partial C_f}{\partial t} = 0 \qquad (3.1.3)$$

Assumption	Justification
A: Both phases present in entire domain	Granular nature of mixture doesn't allow one phase to occupy the entire volume
B: Single pressure for both phases	Grain-to-grain contacts occur randomly. No continuous grain skeleton exists, which could transmit a solid pressure
C: Solid phase as a viscous fluid	The solid material is nearly incompressible, interaction between grains increase with increasing shear
D: The two phases occupy the computational domain completely, without leaving voids	The mass of the flow is assumed to be fully saturated (Section 5.2.2 discusses an extension for the partially saturated case)
E: The volume fractions remain approximately constant during an iteration	No sharp and rapidly moving gradients of volume fractions are present in the two-phase fluid

Table 3.1: Assumptions made in modeling of the two-phase flow.

3.2 Governing equations

The governing equations of two-phase flow are an equation of mass conservation and an equation of momentum conservation for both phases. Furthermore the stress in each phase is given by a constitutive relation and interaction between the two phases is governed by a momentum exchange relation. The model of the two phases smeared over the entire volume occupied by the mixture requires volume averaging of the governing equations of each phase. In the end it should be possible to retrieve the governing equations of a single-phase viscous fluid by summing up the equations of each phase.

The equations of mass and momentum conservation of the phases can be derived by multiplication with the volume fractions and by adding the momentum exchange term to the equation of conservation of momentum. In the following we present a more thorough derivation that can be obtained by volume averaging. Such a derivation is expected to show exactly what assumptions are made if we go from a model with discrete interfaces between the phases to a smeared model, where in each point both phases are present. In the process we assume two distinct phases with well defined interfaces to be blended in a mixture in such a way that the phase interfaces are scattered evenly through the domain. We follow the procedure of volume averaging as it is laid out in the book by Soo [48].

3.2 – Governing equations

3.2.1 Volume averaging of governing equations of two-phase flow

The operation of volume averaging is indicated by triangular brackets $\langle \cdot \rangle$. Volume averaging can be applied to any scalar, vector or tensor quantity ϕ_p belonging to phase p:

$$\langle \phi_p \rangle = \frac{1}{V} \int_{V_p} \phi_p \, dV = C_p \langle \phi_p \rangle^i \quad (3.2.1)$$

V_p is the volume in V that is occupied by phase p, C_p the volume fraction of phase p and V denotes the control volume. The intrinsic average $\langle \phi_p \rangle^i$ is:

$$\langle \phi_p \rangle^i = \frac{1}{V_p} \int_{V_p} \phi_p \, dV \quad (3.2.2)$$

Volume averages of derivatives of a quantity ϕ_p are given by Reynold's transport theorem. The following derivatives of ϕ_p are used:

$$\langle \frac{\partial \phi_p}{\partial t} \rangle = \frac{\partial \langle \phi_p \rangle}{\partial t} - \frac{1}{V} \int_\Gamma \phi_p \mathbf{v}^\Gamma \cdot \mathbf{n}_p \, dS \quad (3.2.3)$$

$$\langle \nabla \phi_p \rangle = \nabla \langle \phi_p \rangle + \frac{1}{V} \int_\Gamma \phi_p \mathbf{n}_p \, dS \quad (3.2.4)$$

$$\langle \nabla \cdot \phi_p \rangle = \nabla \cdot \langle \phi_p \rangle + \frac{1}{V} \int_\Gamma \phi_p \cdot \mathbf{n}_p \, dS \quad (3.2.5)$$

Γ is the interface between phases. $\mathbf{v}^\Gamma$ is the velocity of the interface and $\mathbf{n}_p$ defines the unit normal vector pointing outward from phase p.

Averages of products are expressed as the product of an averaged value with the intrinsic average of the other value:

$$\langle \phi_p \psi_p \rangle = \langle \phi_p \rangle \langle \psi_p \rangle^i \quad (3.2.6)$$

The volume fraction of phase p can be written as a limit as the control volume V tends to zero:

$$C_p = \lim_{V \to 0} \frac{1}{V} \int_{V_p \subset V} 1 \, dV \quad (3.2.7)$$

With these relations we can establish the governing equations of two-phase flow.

3.2.2 Conservation of mass

Volume averaging of Equation 2.3.5 of phase p and application of the rule for averages of products leads to:

$$\langle \frac{\partial \rho_p}{\partial t} \rangle + \langle \nabla \cdot (\rho_p \mathbf{v}_p) \rangle = 0$$

$$\frac{\partial \langle \rho_p \rangle}{\partial t} + \nabla \cdot (\langle \rho_p \rangle^i \langle \mathbf{v}_p \rangle) = -\frac{1}{V} \int_\Gamma \rho_p (\mathbf{v}_p - \mathbf{v}^\Gamma) \cdot \mathbf{n}_p \, dS \quad (3.2.8)$$

The integral on the right-hand side represents the rate of mass generation per unit volume of phase p. Since in our model no mass exchange occurs between the two phases this integral is equal to zero. If we let the control volume V tend to zero as in Equation 3.2.7 we obtain

$$\frac{\partial C_p \rho_p}{\partial t} + \nabla \cdot (C_p \rho_p \mathbf{v}_p) = 0 \qquad (3.2.9)$$

or, in a reference frame attached to the material

$$\frac{D C_p \rho_p}{D t} + C_p \rho_p \nabla \cdot \mathbf{v}_p = 0 \qquad (3.2.10)$$

3.2.3 Conservation of momentum

Volume averaging of the single-phase momentum equation (2.3.12) over phase p gives

$$\langle \rho_p \frac{D \mathbf{v}_p}{D t} \rangle = \langle \nabla \cdot \sigma_p \rangle + \langle \rho_p \mathbf{b} \rangle \qquad (3.2.11)$$

Expanding the inertial term as a product of averages

$$\langle \rho_p \frac{D \mathbf{v}_p}{D t} \rangle = \langle \rho_p \rangle \langle \frac{D \mathbf{v}_p}{D t} \rangle^i \qquad (3.2.12)$$

In order to use Equation 3.2.3 we need to expand the total derivative

$$\begin{aligned}
\langle \frac{D \mathbf{v}_p}{D t} \rangle^i &= \langle \frac{\partial \mathbf{v}_p}{\partial t} \rangle^i + \langle \mathbf{v}_p (\nabla \cdot \mathbf{v}_p) \rangle^i \\
&= \frac{\partial \langle \mathbf{v}_p \rangle^i}{\partial t} - \frac{1}{V} \int_\Gamma \mathbf{v}_p (\mathbf{v}^\Gamma \cdot \mathbf{n}_p) dS \\
&\quad + \langle \mathbf{v}_p \rangle^i \langle \nabla \cdot \mathbf{v}_p \rangle \\
&= \frac{\partial \langle \mathbf{v}_p \rangle^i}{\partial t} - \frac{1}{V} \int_\Gamma \mathbf{v}_p (\mathbf{v}^\Gamma \cdot \mathbf{n}_p) dS \\
&\quad + \langle \mathbf{v}_p \rangle^i \left(\nabla \cdot \langle \mathbf{v}_p \rangle^i + \frac{1}{V} \int_\Gamma \mathbf{v}_p \cdot \mathbf{n}_p dS \right)
\end{aligned} \qquad \begin{aligned} (3.2.13) \\ \\ \\ (3.2.14) \end{aligned}$$

Since $\langle \mathbf{v}_p \rangle^i$ is the convective velocity of phase p, the total derivative of $\langle \mathbf{v}_p \rangle^i$ can be identified as:

$$\begin{aligned}
\langle \frac{D \mathbf{v}_p}{D t} \rangle^i &= \frac{D \langle \mathbf{v}_p \rangle^i}{D t} - \frac{1}{V} \int_\Gamma \mathbf{v}_p (\mathbf{v}^\Gamma \cdot \mathbf{n}_p) dS \\
&\quad + \langle \mathbf{v}_p \rangle^i \frac{1}{V} \int_\Gamma \mathbf{v}_p \cdot \mathbf{n}_p dS
\end{aligned} \qquad (3.2.15)$$

As we let the control volume V, and with it the volume $V_p \subset V$, tend to zero the intrinsic average phase velocity $\langle \mathbf{v}_p \rangle^i$ is equal to $\mathbf{v}_p$. The surface integrals can then be combined:

$$\langle \frac{D \mathbf{v}_p}{D t} \rangle^i = \frac{D \mathbf{v}_p}{D t} - \frac{1}{V} \int_\Gamma \mathbf{v}_p ((\mathbf{v}^\Gamma - \mathbf{v}_p) \cdot \mathbf{n}_p) dS \qquad (3.2.16)$$

3.2 – Governing equations

As for the conservation of mass the integral on the right-hand side reduces to zero because no mass is exchanged between phases.

The volume average of the stress divergence evaluates to the following expression:

$$\langle \nabla \cdot \sigma_p \rangle = \nabla \cdot \langle \sigma_p \rangle + \frac{1}{V} \int_\Gamma \sigma_p \cdot \mathbf{n}_p \, dS \quad (3.2.17)$$

Finally the conservation of momentum of phase p can be written as:

$$C_p \rho_p \frac{D\mathbf{v}_p}{Dt} = \nabla \cdot \langle \sigma_p \rangle + C_p \rho_p \mathbf{b} + \frac{1}{V} \int_\Gamma \sigma_p \cdot \mathbf{n}_p \, dS \quad (3.2.18)$$

The surface integral represents the exchange of momentum between the two phases.

3.2.4 Momentum exchange

As we have seen above a momentum exchange term arises naturally if we apply volume averaging on the momentum conservation equation of one of the phases. This momentum exchange term is expressed in the form of a surface integral over the interface between the two phases. Lets denote the momentum exchange term, transferring momentum from the fluid phase to the solid phase, by $\mathbf{m}_{sf}$:

$$\mathbf{m}_{sf} = \frac{1}{V} \int_\Gamma \sigma_s \cdot \mathbf{n}_s \, dS \quad (3.2.19)$$

From the point of view of the fluid phase the momentum exchange term has to be equal but with opposite sign: $\mathbf{m}_{fs} = -\mathbf{m}_{sf}$.

$$\mathbf{m}_{fs} = -\mathbf{m}_{sf} = \frac{1}{V} \int_\Gamma \sigma_f \cdot \mathbf{n}_f \, dS \quad (3.2.20)$$

In order to verify that this reciprocity is satisfied we have to make sure that the following holds:

$$\sigma_s \cdot \mathbf{n}_s = -\sigma_f \cdot \mathbf{n}_f \quad \text{on } \Gamma \quad (3.2.21)$$

Continuity of the stress at the interface requires $\sigma_s = \sigma_f$. The unit normal vector pointing outward of the fluid phase is equal to the negative of the unit normal vector pointing outward of the solid phase: $\mathbf{n}_s = -\mathbf{n}_f$. Reciprocity between the two momentum exchange terms is therefore satisfied.

As mentioned earlier we don't model any phase interfaces, both phases are smeared over the entire domain. Therefore we need to find a way to describe the momentum exchange that doesn't require an explicit description of an interface. Furthermore due to the smearing over the volume both phases are present in a point in space and their velocities are not equal. Instead of formulating momentum exchange between two spatially separated fluids across an interface we have to find a formulation that relates the phase velocities in one point. To gain insight in the structure of the term let us consider the following analogies:

- Lets imagine two parallel plates separated by a small space h which is filled with a viscous fluid of viscosity μ. If we move one plate with a constant velocity v relative to the other plate then the velocity distribution between the plates is linear. For Newtonian fluids (viscous fluids) the shear stress τ is proportional to the velocity gradient: $\tau = \mu \frac{\partial v}{\partial h}$. The force we have to apply per unit area of the plate in order to maintain the velocity constant can then be expressed by

$$F = \mu \frac{\partial v}{\partial h} h = \mu v \qquad (3.2.22)$$

- Stokes' law says that the drag force F applied to a very small spherical particle of radius r such that it moves through a viscous fluid at constant velocity v is of the form

$$F = 6\pi \mu r v \qquad (3.2.23)$$

μ is the viscosity. For a cloud of particles that is sufficiently dense this formula is no longer valid, the proportionality between F and v however is retained.

Both analogies lead to a linear relationship between the drag force $\mathbf{m}_{sf}$ and the difference in velocities between the two phases. We therefore adopt the following expression:

$$\mathbf{m}_{sf} = -\mathbf{m}_{fs} = K'_{drag}(\mathbf{v}_s - \mathbf{v}_f) \qquad (3.2.24)$$

De la Cruz and Spanos come to the same conclusion in [15] by expanding the surface integral in powers and keeping only the lowest order terms.

An expression for K'_{drag} can be derived by considering a fluid moving past a cloud of particles, which has been done in Soo [48]. This leads to

$$K'_{drag} = \frac{75}{2} \frac{C_s}{(1-C_s)^2} \frac{\mu_f}{a^2} = K_{drag} \frac{C_s}{(1-C_s)^2} \mu_f \qquad (3.2.25)$$

where a is a radius of a particle. $K_{drag} = \frac{75}{2a^2}$ has the unit of $[m^{-2}]$. Its value has to be determined based on an estimation of the particle diameter a. Even if this model is valid only approximately and for an idealized granular phase it can serve as an indication for selecting appropriate values of K_{drag}. For mud-flows it gives values of K_{drag} in the range between 1 for very coarse-grained mixtures and 10^5 for slurry flows with a large fines content. In problems where the volume fractions vary only slightly and the distribution is smooth K'_{drag} can be assumed to be constant.

3.2.5 Constitutive relation

The constitutive relations of the two-phase fluid are derived from the constitutive relation of a single-phase viscous fluid. In order to make sure that the limit case of two phases with exactly

3.2 – Governing equations

the same properties is identical to the situation of a single phase the following relation has to hold:

$$\langle \sigma \rangle = \langle \sigma_s \rangle + \langle \sigma_f \rangle \tag{3.2.26}$$

This relation is satisfied for the following constitutive equations:

$$\langle \sigma_s \rangle = C_s \left(\tau(\mathbf{v}_s) + p\mathbf{I} \right) \tag{3.2.27}$$

$$\langle \sigma_f \rangle = C_f \left(\tau(\mathbf{v}_f) + p\mathbf{I} \right) \tag{3.2.28}$$

The deviatoric stress tensors are defined as

$$\tau(\mathbf{v}_s) = 2\mu_s \left(\dot{\epsilon}(\mathbf{v}_s) - \frac{1}{3}(\nabla \cdot \mathbf{v}_s)\mathbf{I} \right) \tag{3.2.29}$$

$$\tau(\mathbf{v}_f) = 2\mu_f \left(\dot{\epsilon}(\mathbf{v}_f) - \frac{1}{3}(\nabla \cdot \mathbf{v}_f)\mathbf{I} \right) \tag{3.2.30}$$

where μ_s and μ_f are the dynamic viscosities of the solid and the fluid phase. The rates of deformation are given by $\dot{\epsilon}_s = \frac{1}{2}(\nabla \mathbf{v}_s + (\nabla \mathbf{v}_s)^T)$ and $\dot{\epsilon}_f = \frac{1}{2}(\nabla \mathbf{v}_f + (\nabla \mathbf{v}_f)^T)$. With these constitutive relations Equation 3.2.26 becomes

$$\begin{aligned}\langle \sigma \rangle &= \langle \sigma_s \rangle + \langle \sigma_f \rangle \\ &= 2C_s\mu_s \left(\dot{\epsilon}(\mathbf{v}_s) - \frac{1}{3}(\nabla \cdot \mathbf{v}_s)\mathbf{I} \right) + 2C_f\mu_f \left(\dot{\epsilon}(\mathbf{v}_f) - \frac{1}{3}(\nabla \cdot \mathbf{v}_f)\mathbf{I} \right) + p\mathbf{I} \end{aligned} \tag{3.2.31}$$

If we substitute the viscosities μ_s and μ_f by the same viscosity of a single-phase fluid and we assume that both phase velocities are equal, then we obtain the constitutive relation of the single-phase fluid:

$$\begin{aligned}\langle \sigma \rangle &= \langle \sigma_s \rangle + \langle \sigma_f \rangle \\ &= 2(C_s + C_f)\mu \left(\dot{\epsilon}(\mathbf{v}) - \frac{1}{3}(\nabla \cdot \mathbf{v})\mathbf{I} \right) + p\mathbf{I} \\ &= 2\mu \left(\dot{\epsilon}(\mathbf{v}) - \frac{1}{3}(\nabla \cdot \mathbf{v})\mathbf{I} \right) + p\mathbf{I} \end{aligned} \tag{3.2.32}$$

3.2.6 Summary of the initial/boundary value problem

The equations to be solved in the boundary value problem are the two equations of conservation of momentum and one equation of conservation of mass. The equation of conservation of mass is obtained by combining the equations of conservation of mass of each phase and by assuming incompressible behavior of both phases:

$$\frac{\partial \rho_s}{\partial t} = \frac{\partial \rho_f}{\partial t} = 0 \tag{3.2.33}$$

$$\nabla \rho_s = \nabla \rho_f = 0 \tag{3.2.34}$$

Mass conservation of the two-phase mixture is established on a control volume that is fixed in space. In this case we can use Equation 3.2.9 for phase p. Taking into account incompressibility of both phases mass conservation of phase p simplifies to:

$$\frac{\partial C_p}{\partial t} + \nabla \cdot (C_p \mathbf{v}_p) = 0 \qquad (3.2.35)$$

Combining this equation for phase s and for phase f and using the fact that the time derivatives of the volume fractions are related (Equation 3.1.3) yields the following equation of conservation of mass of the two-phase mixture:

$$\nabla \cdot (C_s \mathbf{v}_s) + \nabla \cdot (C_f \mathbf{v}_f) = 0 \qquad (3.2.36)$$

Conservation of momentum can be obtained from combining Equation 3.2.18 with Equations 3.2.27, 3.2.28, 3.2.19 and 3.2.20.

The boundary value problem consists in solving the following equations for velocities and pressure, given the boundary conditions $\mathbf{g}_s$, $\mathbf{g}_f$, $\mathbf{h}_s$, $\mathbf{h}_f$ and the initial conditions $\mathbf{v}_{s,0}$, $\mathbf{v}_{f,0}$ and p_0:

$$C_s \rho_s \frac{D\mathbf{v}_s}{Dt} = \nabla \cdot [C_s(\tau(\mathbf{v}_s) + p\mathbf{I})] + C_s \rho_s \mathbf{b} + \mathbf{m}_{sf} \quad \text{on } \Omega \times]0,T[\qquad (3.2.37)$$

$$C_f \rho_f \frac{D\mathbf{v}_f}{Dt} = \nabla \cdot [C_f(\tau(\mathbf{v}_f) + p\mathbf{I})] + C_f \rho_f \mathbf{b} + \mathbf{m}_{fs} \quad \text{on } \Omega \times]0,T[\qquad (3.2.38)$$

$$0 = \nabla \cdot (C_s \mathbf{v}_s) + \nabla \cdot (C_f \mathbf{v}_f) \quad \text{on } \Omega \times]0,T[\qquad (3.2.39)$$

$$\mathbf{v}_s = \mathbf{g}_s \quad \text{on } \partial\Omega_{g_s} \times]0,T[\qquad (3.2.40)$$

$$\mathbf{v}_f = \mathbf{g}_f \quad \text{on } \partial\Omega_{g_f} \times]0,T[\qquad (3.2.41)$$

$$\sigma_s \cdot \mathbf{n} = \mathbf{h}_s \quad \text{on } \partial\Omega_{h_s} \times]0,T[\qquad (3.2.42)$$

$$\sigma_f \cdot \mathbf{n} = \mathbf{h}_f \quad \text{on } \partial\Omega_{h_f} \times]0,T[\qquad (3.2.43)$$

$$\mathbf{v}_s(t=0) = \mathbf{v}_{s,0} \quad \text{on } \Omega \qquad (3.2.44)$$

$$\mathbf{v}_f(t=0) = \mathbf{v}_{f,0} \quad \text{on } \Omega \qquad (3.2.45)$$

$\partial\Omega_{g_p}$ denotes the part of the boundary on which we impose the displacement $\mathbf{g}_p$, while $\partial\Omega_{h_p}$ denotes the Neumann part, where we impose surface tractions $\mathbf{h}_p$. The index p is a placeholder and can take the values s or f.

3.3 Weak form and stabilization

The global weak form is established as follows: Let $\mathcal{S}_i^s = \{v_i^s \in H^1(\Omega) \mid v_i^s = g_i^s \text{ on } \Gamma_{g_i^s}\}$ and $\mathcal{S}_i^f = \{v_i^f \in H^1(\Omega) \mid v_i^f = g_i^f \text{ on } \Gamma_{g_i^f}\}$ be spaces of solid and fluid trial functions, $\mathcal{V}_i^s = \{w_i^s \in H^1(\Omega) \mid w_i^s = 0 \text{ on } \Gamma_{g_i^s}\}$ and $\mathcal{V}_i^f = \{w_i^f \in H^1(\Omega) \mid w_i^f = 0 \text{ on } \Gamma_{g_i^f}\}$ spaces of solid and fluid test

3.3 – Weak form and stabilization

functions and $\mathcal{P} = \{p \in L^2(\Omega)\}$ a space of both trial and test functions [2]. Then the weak form associated with Equations 3.2.37 to 3.2.45 consists in finding $v_i^s \in \mathcal{S}_i^s$, $v_i^f \in \mathcal{S}_i^f$ and $p \in \mathcal{P}$, such that for all $w_i^s \in \mathcal{V}_i^s$, $w_i^f \in \mathcal{V}_i^f$ and $q \in \mathcal{P}$ the following equation holds:

$$
\begin{aligned}
\mathbf{M}: \quad & \int_\Omega C_s \rho_s \mathbf{w}_s \cdot \dot{\mathbf{v}}_s \, d\Omega + \int_\Omega C_f \rho_f \mathbf{w}_f \cdot \dot{\mathbf{v}}_f \, d\Omega \\
-\mathbf{K}^{\nabla C}: \quad & - \int_\Omega \mathbf{w}_s \nabla C_s \cdot (\tau(\mathbf{v}_s) + p\mathbf{I}) \, d\Omega \\
& - \int_\Omega \mathbf{w}_f \nabla C_f \cdot (\tau(\mathbf{v}_f) + p\mathbf{I}) \, d\Omega \\
-\mathbf{G}^{\nabla C}: \quad & - \int_\Omega \mathbf{w}_s \cdot \nabla C_s \, p \, d\Omega - \int_\Omega \mathbf{w}_f \cdot \nabla C_f \, p \, d\Omega \\
\mathbf{G}: \quad & + \int_\Omega C_s \nabla \cdot \mathbf{w}_s p \, d\Omega + \int_\Omega C_f \nabla \cdot \mathbf{w}_f p \, d\Omega \\
\mathbf{K}: \quad & + \int_\Omega C_s \dot{\epsilon}(\mathbf{w}_s) : c^d(\mathbf{v}_s) \, d\Omega \\
& + \int_\Omega C_f \dot{\epsilon}(\mathbf{w}_f) : c^d(\mathbf{v}_f) \, d\Omega \\
-\mathbf{h}: \quad & - \int_\Gamma \mathbf{w}_s \cdot \mathbf{t} \, d\Gamma - \int_\Gamma \mathbf{w}_f \cdot \mathbf{t} \, d\Gamma \\
-\mathbf{f}: \quad & - \int_\Omega C_s \rho_s \mathbf{w}_s \cdot \mathbf{b} \, d\Omega - \int_\Omega C_f \rho_f \mathbf{w}_f \cdot \mathbf{b} \, d\Omega \\
\mathbf{V}: \quad & + \int_\Omega K_{drag} \mathbf{w}_s \cdot (\mathbf{v}_s - \mathbf{v}_f) \, d\Omega \\
& - \int_\Omega K_{drag} \mathbf{w}_f \cdot (\mathbf{v}_s - \mathbf{v}_f) \, d\Omega \\
\mathbf{G}^T: \quad & + \int_\Omega C_s q \nabla \cdot \mathbf{v}_s \, d\Omega + \int_\Omega C_f q \nabla \cdot \mathbf{v}_f \, d\Omega \\
(\mathbf{G}^{\nabla C})^T: \quad & + \int_\Omega q \nabla C_s \cdot \mathbf{v}_s \, d\Omega + \int_\Omega q \nabla C_f \cdot \mathbf{v}_f \, d\Omega \quad = \quad 0 \quad (3.3.1)
\end{aligned}
$$

The left column identifies the discrete matrices corresponding to each term in the weak form. The semidiscrete matrix form is given below.

3.3.1 Stabilization

The present two-phase formulation of a mixture of two incompressible viscous fluids requires stabilization in order to prevent spurious oscillations in the pressure field. This is due to the choice of equal order interpolation functions, analogously to the single-phase formulation in

---

[2] The letters s and f identify the phases. In general they are used as subscripts, except when there already is an index. In that case they are written as superscripts (Example: The velocity of the solid phase $\mathbf{v}_s = [v_x^s \; v_y^s]^T$.

Section 2.5.3. The stabilization terms to be added to the weak form are the same as previously. The following terms are added to the weak form

$$\underbrace{\sum_{e=1}^{n_{ele}} \tau_e \int_{\Omega^e} \nabla q \cdot \nabla p \, d\Omega}_{\mathbf{S}} - \underbrace{\sum_{e=1}^{n_{ele}} \tau_e \int_{\Omega^e} \nabla q \cdot \mathbf{b} \, d\Omega}_{\mathbf{f}_s} \qquad (3.3.2)$$

where the stabilization parameter τ_e of element e is given by

$$\tau_e = \frac{1}{\sqrt{\left(\frac{2\bar{\rho}}{\Delta t}\right)^2 + \left(\frac{4\bar{\mu}}{\alpha h_e^2}\right)^2}} \qquad (3.3.3)$$

$\bar{\mu} = C_s \mu_s + C_f \mu_f$ is the average viscosity, $\bar{\rho} = C_s \rho_s + C_f \rho_f$ the average density, Δt the time step length and h_e a characteristic element length. For the dimensionless stabilization parameter α the same value as in the single-phase formulation is chosen ($\alpha = 0.5$).

3.3.2 Semidiscrete matrix form

Introducing approximations for the test- and trial functions of the velocities and pressure into the weak form and making use of arbitrariness of the test functions leads to the following semidiscrete matrix form:

$$\begin{bmatrix} \mathbf{M} & 0 \\ 0 & 0 \end{bmatrix} \begin{Bmatrix} \mathbf{a} \\ 0 \end{Bmatrix} + \begin{bmatrix} \mathbf{K} - \mathbf{K}^{\nabla C} + \mathbf{V} & \mathbf{G} - \mathbf{G}^{\nabla C} \\ -\mathbf{G}^T - (\mathbf{G}^{\nabla C})^T & \mathbf{S} \end{bmatrix} \begin{Bmatrix} \mathbf{v} \\ p \end{Bmatrix} = \begin{Bmatrix} \mathbf{h} + \mathbf{f} \\ \mathbf{f}_s \end{Bmatrix} \qquad (3.3.4)$$

Details about the calculation of the elemental arrays and matrices can be found in Appendix A.3. To simplify notation we summarize the above equation in the same form as for the single-phase formulation:

$$\mathbf{M}\mathbf{a} + \mathbf{K}\mathbf{v} = \mathbf{f} \qquad (3.3.5)$$

where the nodal acceleration and velocity vectors are given by

$$\mathbf{a} = [a_x^{s,1} \ a_y^{s,1} \ a_x^{f,1} \ a_y^{f,1} \ 0 \ \cdots \ a_x^{s,I} \ a_y^{s,I} \ a_x^{f,I} \ a_y^{f,I} \ 0 \ \cdots \ a_x^{s,n} \ a_y^{s,n} \ a_x^{f,n} \ a_y^{f,n} \ 0]^T$$
$$\mathbf{v} = [v_x^{s,1} \ v_y^{s,1} \ v_x^{f,1} \ v_y^{f,1} \ p^1 \ \cdots \ v_x^{s,I} \ v_y^{s,I} \ v_x^{f,I} \ v_y^{f,I} \ p^I \ \cdots \ v_x^{s,n} \ v_y^{s,n} \ v_x^{f,n} \ v_y^{f,n} \ p^n]^T$$

The superscript $I \in \{1, \cdots, n\}$ identifies a node.

3.4 Time integration scheme

Time stepping is performed using the generalized trapezoidal algorithm presented in Section 2.4 for the single-phase model. Discretization in time of the semidiscrete matrix form at $t = t_{n+1}$ leads to:

$$\mathbf{M}(\mathbf{x}_{n+1}, C_s^{n+1}, C_f^{n+1})\mathbf{a}_{n+1} + \mathbf{K}(\mathbf{x}_{n+1}, C_s^{n+1}, C_f^{n+1})\mathbf{v}_{n+1} = \mathbf{F}_{n+1}^{ext} \quad (3.4.1)$$

The volume fractions C_s and C_f are recomputed after each iteration. The detailed algorithm allowing to advance the solution to the next time step is given in Table 3.2.

The mesh update algorithm for the two-phase formulation is described next.

3.4.1 Mesh update

Table 3.3 summarizes the algorithm, which is very similar to the single-phase algorithm in the previous chapter (Table 2.2). The main difference resides in point 1, where the Lagrangian update produces two updated nodes. This is addressed in Section 3.4.1.1. Points 2 and 3 are identical in the single-phase algorithm. Note that re-meshing is now required in order to maintain a constant number of nodes. The re-mapping of nodal variables in point 4 is also slightly different. The phase velocities are mapped from the corresponding deformed mesh of the phase onto the new mesh. The pressure is mapped from the union of all updated nodes of the two phases onto the new mesh.

3.4.1.1 Lagrangian update

The updating of the spatial coordinates of the nodes is slightly different from the single-phase case. The update is still performed for the predictor step, applying the displacement increment $\Delta \tilde{\mathbf{d}}_{n+1}$, and for the corrector step, using $\Delta \mathbf{d}_{n+1}^{i+1}$. For clarity we explain the update procedure for a general update using an increment $\Delta \mathbf{d}$, updating a coordinate $\mathbf{x}^0$ to $\mathbf{x}^1$. The update yields the new nodal coordinates

$$\mathbf{x}_s^1 = \mathbf{x}^0 + \Delta \mathbf{d}_s \quad (3.4.2)$$
$$\mathbf{x}_f^1 = \mathbf{x}^0 + \Delta \mathbf{d}_f \quad (3.4.3)$$

$\mathbf{x}^0$ is the previous spatial coordinate of the node, which is identical for both phases, and $\mathbf{x}_s^1$ and $\mathbf{x}_f^1$ are the new spatial coordinates of the solid and the fluid material points.

For nodes that are part of the boundary before the Lagrangian update a slightly different update strategy is adopted. A unique displacement increment for both phases is applied to

1. At t_{n+1} do:

   Initialize the iteration counter: $i = 0$

   Predictor phase:
   $$\mathbf{x}_{n+1}^{i=0} = \mathbf{x}_n + \Delta \tilde{\mathbf{d}}_{n+1}$$

   - Mesh update
   - Compute volume fractions $C_s(\mathbf{x}_{n+1}^{i+1})$ and $C_f(\mathbf{x}_{n+1}^{i+1})$ (Section 3.6)

   $$\mathbf{v}_{n+1}^{i=0} = \tilde{\mathbf{v}}_{n+1}$$
   $$\mathbf{a}_{n+1}^{i=0} = 0$$

2. Compute the residual force, the tangent stiffness matrix and solve the linear system of equations:
   $$\Delta \mathbf{F} = \mathbf{F}_{n+1}^{ext} - \mathbf{N}(\mathbf{a}_{n+1}^i, \mathbf{v}_{n+1}^i, \mathbf{x}_{n+1}^i, C_s(\mathbf{x}_{n+1}^i), C_f(\mathbf{x}_{n+1}^i))$$
   $$\mathbf{K}^* = \frac{1}{\Delta t \gamma} \mathbf{M}(\mathbf{x}_{n+1}^i, C_s(\mathbf{x}_{n+1}^i), C_f(\mathbf{x}_{n+1}^i)) + \mathbf{K}(\mathbf{x}_{n+1}^i, C_s(\mathbf{x}_{n+1}^i), C_f(\mathbf{x}_{n+1}^i))$$
   $$\mathbf{K}^* \Delta \mathbf{v} = \Delta \mathbf{F}$$

3. Corrector phase:
   $$\mathbf{v}_{n+1}^{i+1} = \mathbf{v}_{n+1}^i + \Delta \mathbf{v}$$
   $$\mathbf{a}_{n+1}^{i+1} = \frac{1}{\Delta t \gamma}(\mathbf{v}_{n+1}^{i+1} - \tilde{\mathbf{v}}_{n+1})$$
   $$\mathbf{x}_{n+1}^{i+1} = \mathbf{x}_{n+1}^i + \Delta \mathbf{d}_{n+1}^{i+1} = \mathbf{x}_{n+1}^i + \frac{\Delta t \beta}{\gamma} \Delta \mathbf{v}$$

   - Mesh update
   - Compute volume fractions $C_s(\mathbf{x}_{n+1}^{i+1})$ and $C_f(\mathbf{x}_{n+1}^{i-1})$ (Section 3.6)

4. Test if computation has converged: If $|\Delta \mathbf{F}| < C \in \mathbb{R}$, go to 1. (step $n = n + 1$). Else go to 2. (iteration $i = i + 1$).

Table 3.2: Generalized trapezoidal algorithm for two-phase formulation.

3.4 – Time integration scheme

1. Update the spatial coordinates of the nodes

    (a) Predictor step:

    - Interior nodes:

    $$\tilde{\mathbf{x}}_s^{n+1} = \mathbf{x}_s^{n+1,i=0} = \mathbf{x}^n + \Delta \tilde{\mathbf{d}}_s^{n+1} \qquad (1a)$$

    $$\tilde{\mathbf{x}}_f^{n+1} = \mathbf{x}_f^{n+1,i=0} = \mathbf{x}^n + \Delta \tilde{\mathbf{d}}_f^{n+1} \qquad (1b)$$

    - Boundary nodes:

    $$\tilde{\mathbf{x}}_m^{n+1} = \mathbf{x}_m^{n+1,i=0} = \mathbf{x}^n + \Delta \tilde{\mathbf{d}}_m^{n+1}$$
    $$= \mathbf{x}^n + C_s \Delta \tilde{\mathbf{d}}_s^{n+1} + C_f \Delta \tilde{\mathbf{d}}_f^{n+1} \qquad (1c)$$

    with

    $$\Delta \tilde{\mathbf{d}}_s^{n+1} = \Delta t \mathbf{v}_s^n + (0.5 - \beta) \Delta t^2 \mathbf{a}_s^n$$
    $$\Delta \tilde{\mathbf{d}}_f^{n+1} = \Delta t \mathbf{v}_f^n + (0.5 - \beta) \Delta t^2 \mathbf{a}_f^n$$

    (b) Corrector step:

    - Interior nodes:

    $$\mathbf{x}_s^{n+1,i+1} = \mathbf{x}_{n+1}^i + \Delta \mathbf{d}_s^{n+1,i+1} = \mathbf{x}_s^{n+1,i} + \frac{\Delta t \beta}{\gamma} \Delta \mathbf{v}_s \qquad (2a)$$

    $$\mathbf{x}_f^{n+1,i+1} = \mathbf{x}_{n+1}^i + \Delta \mathbf{d}_f^{n+1,i+1} = \mathbf{x}_f^{n+1,i} + \frac{\Delta t \beta}{\gamma} \Delta \mathbf{v}_f \qquad (2b)$$

    - Boundary nodes:

    $$\mathbf{x}_m^{n+1,i+1} = \mathbf{x}_{n+1}^i + \Delta \mathbf{d}_m^{n+1,i+1}$$
    $$= \mathbf{x}_{n+1}^i + C_s \Delta \mathbf{d}_s^{n+1,i+1} + C_f \Delta \mathbf{d}_f^{n+1,i+1} \qquad (2c)$$

    with

    $$\Delta \mathbf{d}_s^{n+1,i+1} = \frac{\Delta t \beta}{\gamma} \Delta \mathbf{v}_s \quad \text{and} \quad \Delta \mathbf{d}_f^{n+1,i+1} = \frac{\Delta t \beta}{\gamma} \Delta \mathbf{v}_f$$

2. Find the boundary of the fluid, using the α-shape method (see Appendix A.2)
3. Re-mesh inside the boundary
4. Re-map the nodal variables on the new mesh (see Section 2.5.5)

Table 3.3: Mesh update algorithm for two-phase formulation.

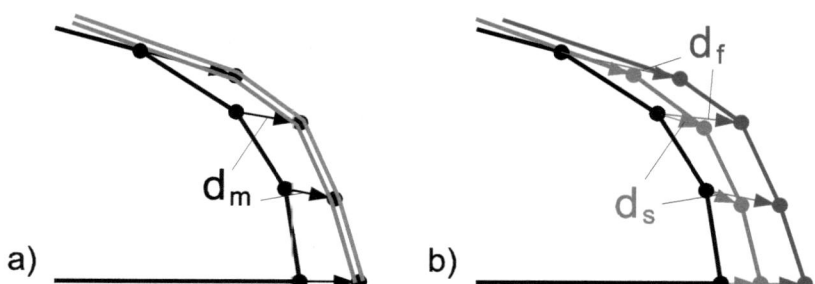

Figure 3.1: Lagrangian update of nodes on the boundary: a) Modified update using displacement of mixture d_m, b) Strict update using displacements of phases d_s and d_f.

the material points of the solid and the fluid phase in order to update the boundary to a well-defined position (Figure 3.1 a). The unique displacement of the boundary is given by the displacement of the mixture:

$$\mathbf{d}_m = C_s \mathbf{d}_s + C_f \mathbf{d}_f \qquad (3.4.4)$$

Remark 3.4.1 *In this work we tried other approaches that relax the hypothesis of presence of both phases to the benefit of applying the real displacements of the phases, according to Figure 3.1 b). This brings up several problems. The updated configuration is not consistent with the assumption that the volume fractions remain approximately constant during an iteration. Since the velocities at the boundary are obtained by assuming volume fractions between 0 and 1, the single-phase domain after the update should only contain that same volume fraction of the single-phase material. The complement would have to be void in order to conserve mass.*

Remark 3.4.2 *Using the displacement increment of the mixture for the update of the boundary satisfies Equation 3.2.36 and therefore conserves the volume of the mixture*

$$\nabla \cdot \mathbf{v}_m = \nabla \cdot (C_s \mathbf{v}_s + C_f \mathbf{v}_f) = \nabla \cdot (C_s \mathbf{v}_s) + \nabla \cdot (C_f \mathbf{v}_f) = 0 \qquad (3.4.5)$$

3.5 Spatial discretization

The computational domain is discretized in the same way as for the single-phase formulation. Each node has 5 degrees of freedom, two components for the velocity of each phase and one pressure. The approximation is defined on linear finite element triangles, which are obtained from a Delaunay triangulation.

3.6 Computation of volume fractions

Computation of volume fractions is done at the beginning of each time step and after each iteration, based on the computed velocity fields v_s and v_f. The problem can be stated as follows: *Using the set of nodes before the update and the nodal sets of the fluid and the solid phases after the update, the volume fractions are to be computed on the new, re-zoned nodal set.* While Equation 3.2.36 ensures conservation of the global volume the algorithm for computing the volume fractions is responsible for conserving the mass of each phase (or volume in the case of incompressible media). Furthermore the method should yield smooth volume fraction fields. Several methods have been tried out, the one retained is the one that best respects the above requirements. In the following some steps of the process that led to the current method are explained.

3.6.1 Background

De-coupling of the computation of the volume fractions from the computation of the velocities and the pressure is justified by the assumption that variations of volume fractions during an iteration are small. The volume fractions can therefore be updated a posteriori based on the positions of the nodes while being kept constant during the computation of the primary unknowns. The idea is that the Lagrangian update of the nodes creates a new distribution of the phases, where all the information necessary to compute volume fractions is available at the updated nodes. Computation of volume fractions basically consists of evaluating the local density of each phase. In the following we investigate several ideas.

The first approach considers the two sets of nodes with a triangulation for each phase separately. The phase mass (or volume) of each node is distributed onto the nodes of the containing triangle of the new mesh according to the barycentric coordinates. This is schematically illustrated in Figure 3.2 a). In this process the total mass of both phases is deposited completely on the nodes of the new mesh. The volume fractions are then computed by dividing the phase volume in each node by the sum of the volumes of both phases. This method conserves mass exactly, however the volumes of material deposited on the new nodes do not necessarily fit the local volume available to the nodes. This method turned out to be too sensitive to the mesh and yielded unstable results.

Similar methods can be imagined, where a large number of particles is pushed through the mesh according to the velocity interpolated at their initial position. The idea is illustrated in

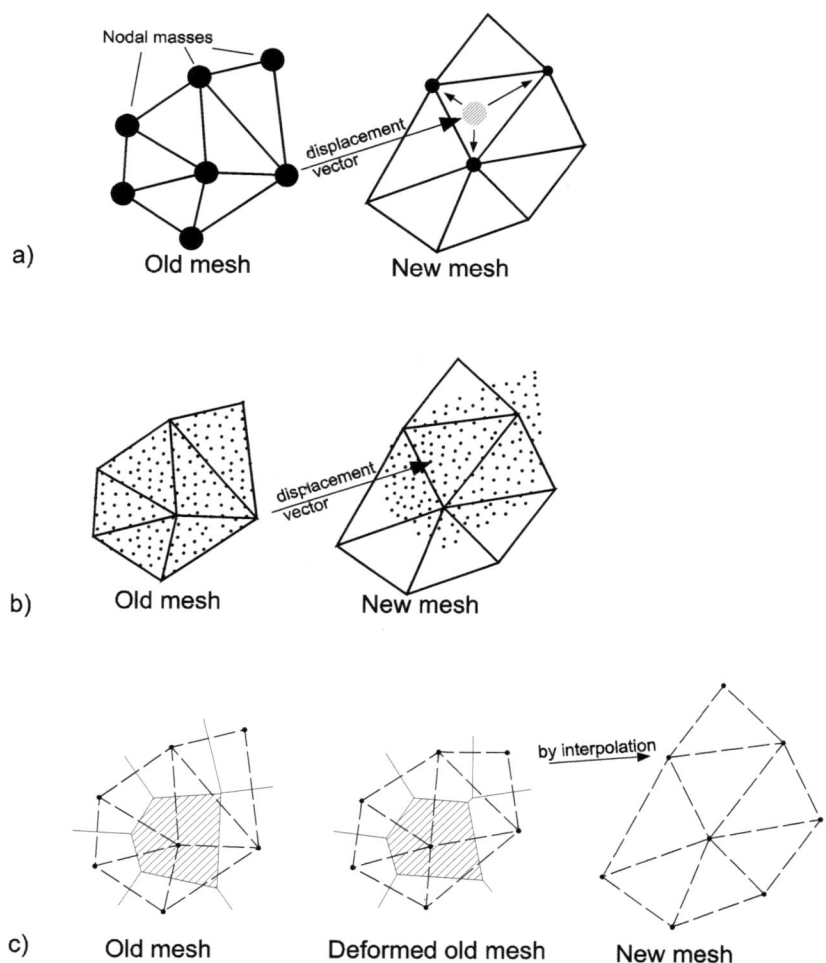

Figure 3.2: Methods for computing volume fractions: a) Convection of nodal masses, b) Particle pushing or Particle-In-Cell, c) Compaction of Voronoi cells.

3.6 – Computation of volume fractions

Figure 3.2 b). The mass of the particles is then attributed to the containing element of the new mesh, or it is deposited onto the nodes as in the method above. In this method mass is also conserved exactly, however the computational cost of performing a large number of point-in-triangle searches can become prohibitively high. Such methods are sometimes referred to as *particle pushing* or *Particle-In-Cell* (PIC) methods and are commonly used in the field of plasma physics. For an example of a PIC-method for computing multi-phase flows Andrews et al. [1] is noted. In their paper the authors present a de-coupled method where the governing equations of the fluid phase are solved on an Eulerian grid and the particle phase is treated separately in a Lagrangian framework.

Finally Figure 3.2 c) shows a method that attempts to capture densification or dedensification of a cloud of nodes, using Voronoi cells as nodal control volumes. The volume of material of one phase in the control volume is assumed to be constant before and after the Lagrangian update. By computing the change in total volume of the Voronoi cell new volume fractions can be computed on the deformed mesh of both phases. In a re-mapping step these volume fractions have to be interpolated on the nodes of the new mesh. The interpolated volume fractions then need to be regularized in order to satisfy $C_s + C_f = 1$. The experience with such techniques however has shown strong mesh dependence, due to which the resulting volume fraction fields are not smooth and often oscillate.

Remark 3.6.1 *Another approach for computing volume fractions consists in using a differential equation, the conservation of mass of a single phase, and computing volume fractions using a time-stepping algorithm. The results of an initial implementation using an Eulerian reference frame were not satisfying. A Lagrangian approach might be more adequate. The main ideas are outlined in Appendix A.4.*

3.6.2 Current Method

In the methods illustrated above mesh dependence is the biggest problem, leading to volume fraction distributions that are not smooth. The current approach therefore attempts to obtain an estimate of the densification of nodal masses, that is independent of any nodal connectivity. Instead of evaluating the densification around a node by exactly computing volumes of Voronoi cells, we evaluate the distances between the node and its neighbors. Weighting the volume fractions of the neighbors with the distance to the node of interest gives an estimation of the local density of a phase. The method is described in the following.

$A_i^{0,p}$ denotes the approximation of the local density of phase p at node i before the Lagrangian update, $A_i^{1,p}$ is the same value after the update, i.e. on the deformed configuration. The local densities $A_i^{0,p}$ and $A_i^{1,p}$ are evaluated using a linear hat-function of radius R, centered at node i:

$$A_i^{0,p} = \sum_j (R - d_{ij}^0) C_j^{0,p} \qquad (3.6.1)$$

$$A_i^{1,p} = \sum_j (R - d_{ij}^{1,p}) C_j^{0,p} \qquad (3.6.2)$$

where $C_j^{0,p}$ is the volume fraction of phase p at node j, before the update. d_{ij}^0 and $d_{ij}^{1,p}$ are the distances between nodes i and j in the undeformed and in the deformed configuration of phase p. The method is illustrated in Figure 3.3. Since the approximated local densities are computed

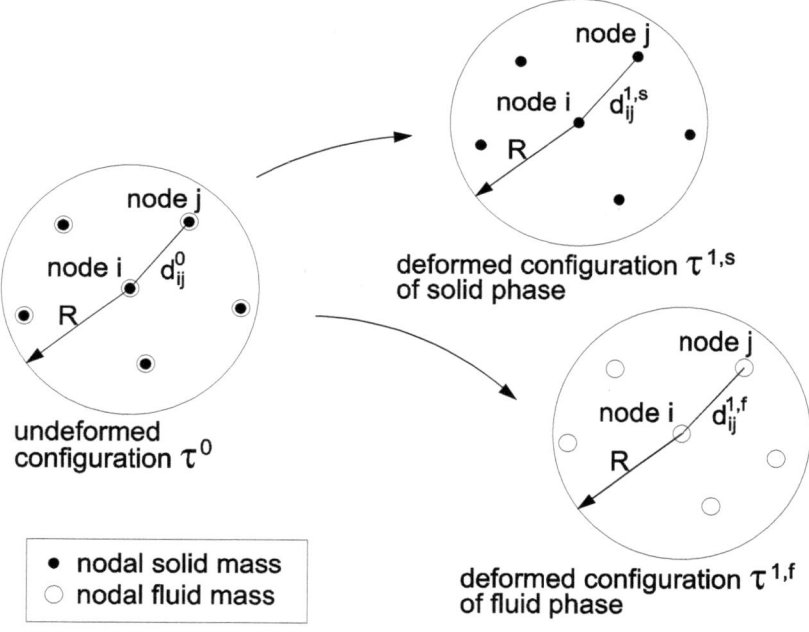

Figure 3.3: Illustration of the method for computing volume fractions on the deformed configurations of both phases.

on the nodes of the old mesh, they have to be interpolated onto the re-zoned nodes of the new mesh. In each node I of the new mesh the volume fraction of the solid phase is computed

3.6 – Computation of volume fractions

according to

$$C_I^{1,s} = C_I^{0,s} + \frac{A_I^{1,s}}{(A_I^{1,s} + A_I^{1,f})} - \frac{A_I^{0,s}}{(A_I^{0,s} + A_I^{0,f})} \qquad (3.6.3)$$

$$C_I^{1,f} = C_I^{0,f} + \frac{A_I^{1,f}}{(A_I^{1,s} + A_I^{1,f})} - \frac{A_I^{0,f}}{(A_I^{0,s} + A_I^{0,f})} \qquad (3.6.4)$$

In a last step, the volume fractions are multiplied by a correction factor such that the total volume occupied by either phase doesn't change. This correction factor is defined for the entire domain as

$$\Lambda^p = \frac{\text{mass of phase } p \text{ after deformation}}{\text{initial (exact) mass}} \qquad (3.6.5)$$

The method is summarized in Table 3.4.

---

1. For each node i of the **old mesh**, do

    For each phase $p = s, f$, do

    1.1 For each node j within a distance R, evaluate distance d_{ij}^0 to node i in the undeformed configuration and compute $A_i^{0,p} = \sum_j (R - d_{ij}^0) C_j^{0,p}$

    1.2 For each node j within a distance R, evaluate distance $d_{ij}^{1,p}$ to node i in the deformed configuration and compute $A_i^{1,p} = \sum_j (R - d_{ij}^{1,p}) C_j^{0,p}$

    1.3 Interpolate $A_i^{0,p}$ and $A_i^{1,p}$ onto the new re-zoned mesh

2. For each node I of the **new mesh**, compute new volume fractions (Equations 3.6.3 and 3.6.4)

3. Correct volume fractions: $C_I^p = C_I^{1,p} \Lambda^p$ (Equation 3.6.5)

Table 3.4: Algorithm for computing volume fractions.

Remark 3.6.2 *The radius R in Equations 3.6.1 and 3.6.2 is a parameter that has to be chosen by the user. It has to be small in order to limit numerical diffusion of sharp gradients, but it cannot be too small in order to still produce smooth results. Appropriate values are in the range between $0.6h$ and $0.8h$,* where h is the average spacing between nodes. Tests showed that in this range varying the radius R has very little influence on the resulting volume fraction fields.

3.6.2.1 Analysis of the performance of the method for computing volume fractions

The method presented above is capable of computing volume fractions based on a neighborhood of nodes. In order to examine the influence of the mesh and the length of the time step, thus the displacement, on the resulting volume fraction field, we impose a predefined motion to the nodes of the mesh. This way we eliminate the physics of two-phase flow from the method which allows us to focus on the purely algorithmic computation of volume fractions. We perform two tests on the unit squares shown in Figure 3.4 a) and b). In both tests the motion

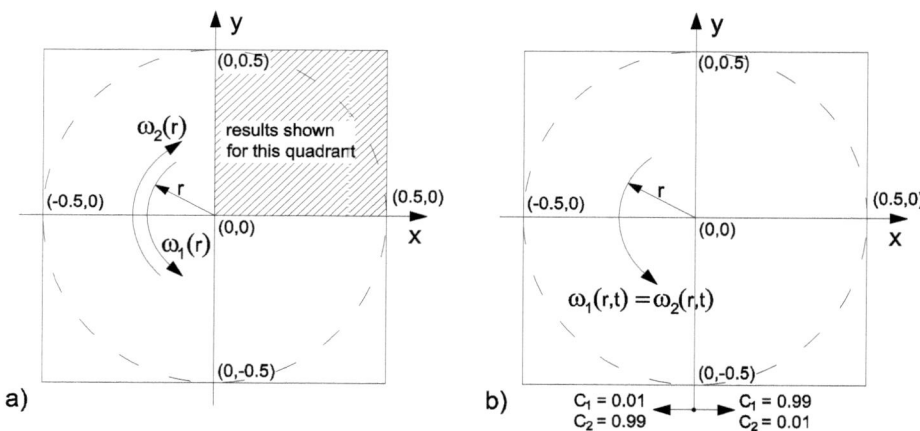

Figure 3.4: Analysis of the method for computing volume fractions. Illustration of the performed tests.

imposed to the nodes of the mesh corresponds to a vortex. The nodes rotate around the origin with an angular velocity of $\omega(r)$, given by

$$\omega(r) = \begin{cases} (0.5 - r)^2 & \forall\, r < 0.5 \\ 0 & \forall\, r \geq 0.5 \end{cases} \tag{3.6.6}$$

3.6 – Computation of volume fractions

where r is the distance from the origin. The new coordinates (x', y') of a node with initial coordinates (x, y) can be computed as

$$\begin{Bmatrix} x' \\ y' \end{Bmatrix} = \begin{bmatrix} \cos(\omega) & -\sin(\omega) \\ \sin(\omega) & \cos(\omega) \end{bmatrix} \begin{Bmatrix} x \\ y \end{Bmatrix} \qquad (3.6.7)$$

In these tests we use a fixed structured mesh on which the volume fractions are interpolated after each increment of motion. In other words, the re-zoned mesh is always identical to the initial undeformed mesh.

Uniform initial volume fraction distribution. In the first test, shown in Figure 3.4 a), the entire square is filled with a uniform mixture of phases 1 and 2. The initial volume fractions are $C_1 = C_2 = 0.5$. On phase 1 a rotation of $\omega_1(r) = \omega(r)$ about the origin is applied, while on phase 2 a rotation of $\omega_2(r) = -\omega(r)$ is imposed. The test is performed using different time step sizes: $\Delta t = \{1, 0.5, 0.2, 0.05, 0.02, 0.01\}$. The total time analyzed is equal to 1 for all computations. In Figure 3.5 the results are shown on the upper right quadrant of the unit square. The amplitude

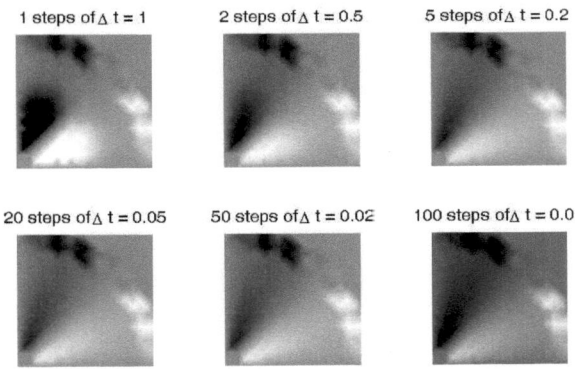

Figure 3.5: Deviation from initial distribution of volume fractions.

of the deviation from the initial volume fraction is about $2.5 \cdot 10^{-4}$ for all computations. Because the patterns are almost identical we can conclude that the error in the volume fractions does not depend on the number of time steps, it depends only on the final displacement a node undergoes with respect to its neighbors. The computation of volume fractions at the end of each iteration does not introduce a constant error.

Two vertically separated phases. In this test, illustrated in Figure 3.4 b), the rotation imposed to both phases is

$$\omega_1(r) = \omega_2(r) = \begin{cases} \omega(r) & \forall\, t < 20 \\ -\omega(r) & \forall\, t \geq 20 \end{cases} \quad (3.6.8)$$

The vortex motion is imposed in 20 steps of length $\Delta t = 1$, then the motion is reversed for another 20 steps. The final positions of all material points are identical with the initial positions. The goal is to see how well the straight vertical line separating the two phases is preserved during the motion and at the end of the computation. The initial volume fractions are $C_1 = 0.99$ and $C_2 = 0.01$ on the right half and $C_1 = 0.01$ and $C_2 = 0.99$ on the left half. The computations are performed on four meshes consisting of 221, 613, 1301 and 3281 nodes. Figure 3.6 shows the volume fractions after 20 steps and after 40 steps. The results show that the accuracy with

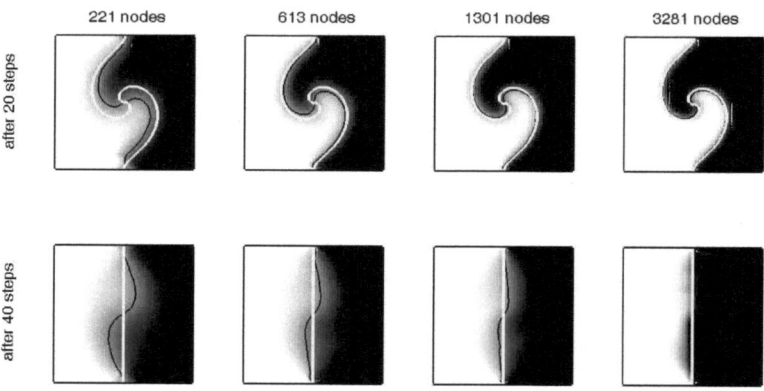

Figure 3.6: Distribution of volume fractions for imposed vortex motion. The white line indicates the exact solution, the thin black line indicates the contour line of $C_1 = C_2 = 0.5$ of the numerical result.

which the separation between the two regions is captured increases as the mesh is refined. The gradient of volume fraction becomes steeper near the separation while at the same time the contour line $C_1 = C_2 = 0.5$ gets closer to the exact separation line.

3.7 Numerical tests

The two-phase model is verified and validated according to the program of tests shown in Table 3.5. The tests have to verify that the implementation of the model accurately represents the conceptual description. Some of these aspects are quantifiable, such as conservation of mass or energy and speed of sedimentation. Other aspects are to be verified qualitatively, such as the ability of the denser phase to settle within the mixture, or the change in behavior as a parameter is varied. Validation of the two-phase model is mainly done in a qualitative manner. This due to the lack of suitable experimental results of flow events that can be represented by a two-dimensional model. Also the present computational method primarily provides a framework for later implementation of constitutive behaviors that more closely represent debris flows or mudflows. For an in-depth review of verification and validation of numerical models the reader is referred to Oberkampf et al. [39].

Two-phase tests	Aspect to be verified	Comparison with
Laminar flow	Equivalence with single-phase model	Single-phase
Flow over backward-facing step	Equivalence with single-phase model	Single-phase
Flow over backward-facing step with free surface	Stability of free surface	Qualitatively
Sedimentation	Drag force, constitutive model	Soo [48]
Vertical separation	Sharp gradients of volume fractions	Qualitatively
Heavy drop	Sharp gradients of volume fractions, free surface	Qualitatively
Dam break	Effect of varying drag force coefficient	Qualitatively
Dam break on inclined slope impacting on obstacle	Force acting on obstacle	Qualitatively, hand calculation

Table 3.5: Verification tests for two-phase model.

3.7.1 Equivalence between two-phase and single-phase fluid

The two-phase formulation is based on viscous flow of a single phase fluid. Therefore the behavior of the single-phase fluid has to be recovered as the material properties density and viscosity of both phases of the two-phase fluid are set equal to the parameters of the single-phase fluid. Two numerical tests, stratified flow as in Section 2.6.3 and flow over a backward-facing step as in Section 2.6.4 are performed with different (but constant) volume fractions and drag coefficients. Drag coefficients K_{drag} between 0 and 10'000 and solid volume fractions C_s in the range between 0.5 and 0.8 have been tested. The L_2-norms of the difference in velocity ($||\mathbf{v}^{tp} - \mathbf{v}^{sp}||_{L_2}$, sp for single-phase and tp for two-phase) and pressure ($||p^{tp} - p^{sp}||_{L_2}$) obtained for all the tests are equal to machine-zero for all time steps. We can thus claim that the two-phase formulation is equivalent to the corresponding single-phase formulation, when both phases have identical properties. In this case the two-phase formulation also converges quadratically to the exact solution, as has been shown for the single-phase formulation in Section 2.6.3.

3.7.2 Sedimentation

Sedimentation of a phase of solid particles in a viscous fluid is a problem of great interest for many industrial processes. In the context of debris flows it is an important test for the interaction behavior between the two phases. This test serves as a verification that the drag force term acts as intended. We compare our results with the analytical solution given by Soo [48]. His formulation is slightly different from ours, therefore some adjustments need to be made. In his solution, Soo neglects diffusion. The conservation of momentum of the solid and the fluid phase are, using our notation and re-arranging some terms

$$C_s \rho_s \frac{Dv_s}{Dt} = -C_s \frac{\partial p}{\partial z} - C_s (\rho_s - \rho_f) g + C_s \rho_s F (v_f - v_s) \qquad (3.7.1)$$

$$C_f \rho_f \frac{Dv_f}{Dt} = -C_f \frac{\partial p}{\partial z} - \rho_f g - C_s \rho_s F (v_f - v_s) \qquad (3.7.2)$$

Using the same one-dimensional form in vertical direction Equations 3.2.37 and 3.2.38 become, using the same convention for pressure

$$C_s \rho_s \frac{Dv_s}{Dt} = \frac{\partial \left[C_s (\sigma_{zz}^d (v_s) - p) \right]}{\partial z} - C_s \rho_s \mathbf{g} + \mathbf{m}_{sf} \qquad (3.7.3)$$

$$C_f \rho_f \frac{Dv_f}{Dt} = \frac{\partial \left[C_f (\sigma_{zz}^d (v_f) - p) \right]}{\partial z} - C_f \rho_f \mathbf{g} - \mathbf{m}_{sf} \qquad (3.7.4)$$

Remark 3.7.1 *The body force is defined slightly differently in the two formulations. Soo defines it as a buoyant density. Furthermore, in order to be able to obtain an analytic solution, he assumes that $\rho_f \ll \rho_s$. Under this assumption the two forms are identical. We use $\rho_s = 10$ and $\rho_f = 0.001$.*

Remark 3.7.2 *By neglecting diffusion Soo sets σ_{zz}^d equal to zero. For numerical reasons we cannot do this in our model, however by choosing small values of viscosity this is approximately true. In this analysis we use $\mu_s = \mu_f = 0.01$.*

3.7 – Numerical tests

Considering the two remarks above, the equations used by Soo are almost identical to Equations 3.7.3 and 3.7.4. The only remaining difference is the term involving the gradient of the volume fraction. Such a $p\nabla C_s$-term is present in the model by Soo. This term acts as a force against the separation of the phases. In order to verify the implementation we omit the volume-fraction gradient for this test. The analysis is performed on a rectangular container of height $h = 1$ and width $d = 0.15$. At time $t = 0$ the volume is filled with a homogeneous mixture with a solid volume fraction of $C_s = 0.2$. The maximum volume fraction is limited to $C_s^{max} = 0.5$, which represents the void ratio of a granular material after sedimentation is completed. In order to stop the sedimentation process when C_s^{max} is reached we simply increase the viscosities of both phases by a factor of 10^6.

Remark 3.7.3 *Increasing the viscosities when C_s^{max} is reached can be considered a simple form of a non-Newtonian constitutive model. In this case the constitutive relations (Equations 3.2.27 and 3.2.28 are modified as*

$$\tau(\mathbf{v}_s) = 2\mu_s K(C_s)\left(\dot{\epsilon}(\mathbf{v}_s) - \frac{1}{3}(\nabla \cdot \mathbf{v}_s)\mathbf{I}\right)$$

$$\tau(\mathbf{v}_f) = 2\mu_f K(C_s)\left(\dot{\epsilon}(\mathbf{v}_f) - \frac{1}{3}(\nabla \cdot \mathbf{v}_f)\mathbf{I}\right)$$

where $K(C_s)$ is a penalty coefficient defined as

$$K(C_s) = \begin{cases} 1 & \text{if } C_s \leq C_s^{max} - r \\ \lambda & \text{if } C_s > C_s^{max} - r \end{cases}$$

The penalty value λ is applied as the volume fraction C_s is approaching C_s^{max} by less than a value r. Alternatively the penalty can be increased linearly between $C_s^{max} - r$ and C_s^{max}.

The minimum solid volume fraction is limited to $C_s^{min} = 0.01$. The following definition of the drag force coefficient is used

$$K'_{drag} = F'C_s\rho_s \qquad (3.7.5)$$

Analogously to Soo, $F' = 10$ is set. In Figure 3.7 we show a comparison of the analytical results by Soo with our results, along with the mesh used in the computation. The figure shows the solid volume fraction along the vertical axis, versus time. The results match very well. The time for complete separation of phases is reproduced almost exactly. In the numerical results the separation between the zones A, B and C, where the analytical solution assumes constant volume fractions, is more diffuse due to the physical diffusion, but also due to some numerical

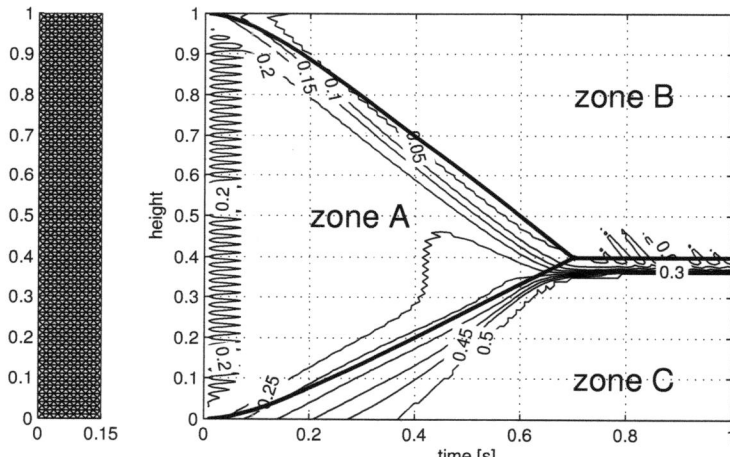

Figure 3.7: Contours of solid volume fraction as a function of vertical coordinate and time for the sedimentation problem, using the formulation by Soo. The thick line indicates the delimitation of three separate zones given by the analytical solution: Zone A: $C_s = 0.2$, Zone B: $C_s = 0$, Zone C: $C_s = 0.5$.

3.7 – Numerical tests

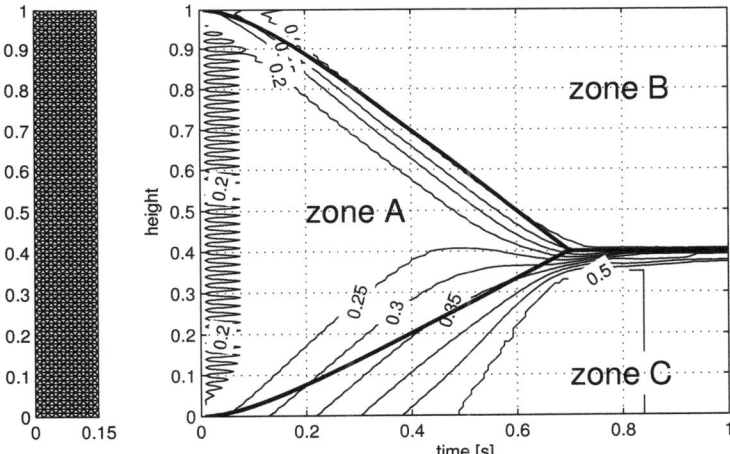

Figure 3.8: Contours of solid volume fraction as a function of vertical coordinate and time for the sedimentation problem, using the original formulation with the $p\nabla C_s$-term. The thick line indicates the delimitation of three separate zones given by the analytical solution: Zone A: $C_s = 0.2$, Zone B: $C_s = 0$, Zone C: $C_s = 0.5$.

diffusion. Still three areas of almost constant volume fraction can clearly been identified. For comparison we also show the same result using the original formulation, including the $p\nabla C_s$-term (Figure 3.8). The main difference between the two results is in the separation between zones A and C, where the formulation omitting the $p\nabla C_s$-term has a much sharper interface. We also note that including the $p\nabla C_s$-term yields slightly smoother results.

In order to show that the separation between the two phases becomes sharper as the mesh is refined we plot the vertical profile of solid volume fractions in Figure 3.9. We use six different meshes with 17 nodes in the coarsest mesh and 859 in the finest. The spacing of nodes in vertical direction was 0.1, 0.05, 0.0333, 0.025, 0.0147 and 0.01 for the finest mesh. The time steps were also adapted to the mesh size in such a way that the maximum differential displacement per time step, divided by the vertical mesh spacing, $|v_s - v_f|\Delta t/\Delta h$, doesn't exceed a value of 0.9. From the coarsest to the finest mesh, Δt was chosen to be 0.08, 0.04, 0.025, 0.02, 0.0125 and 0.008. The profile is given at $t = 0.7$ and $t = 2$. Figure 3.9 shows that on sufficiently fine meshes very steep gradients can be reproduced. The corresponding result, this time including the $p\nabla C_s$-

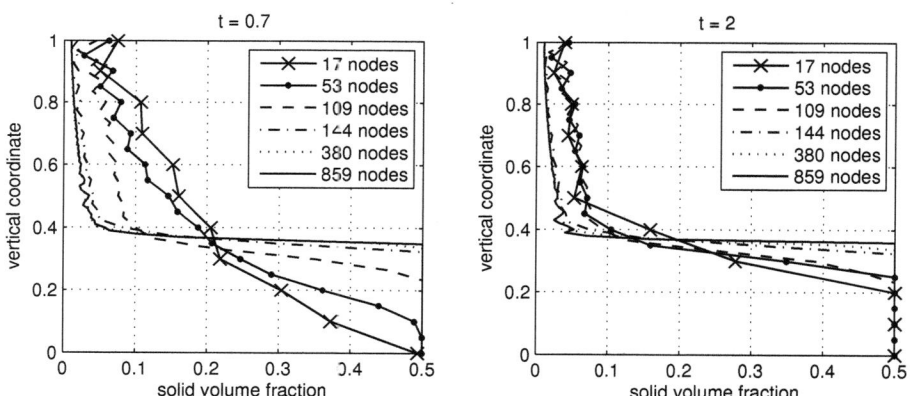

Figure 3.9: Vertical profiles of solid volume fraction at $t = 0.7$ and $t = 2$. Formulation without volume-fraction-gradient term.

term shows in Figure 3.10 again slightly smoother results. The presence of strong gradients of volume fraction can in this formulation however lead to instabilities, as can be seen in the plot for $t = 2$. Weather or not the term has to be included depends on the nature of the problem to be analyzed. Calibration with experiments are required, as it is customary for parameters of a constitutive model.

3.7 – Numerical tests

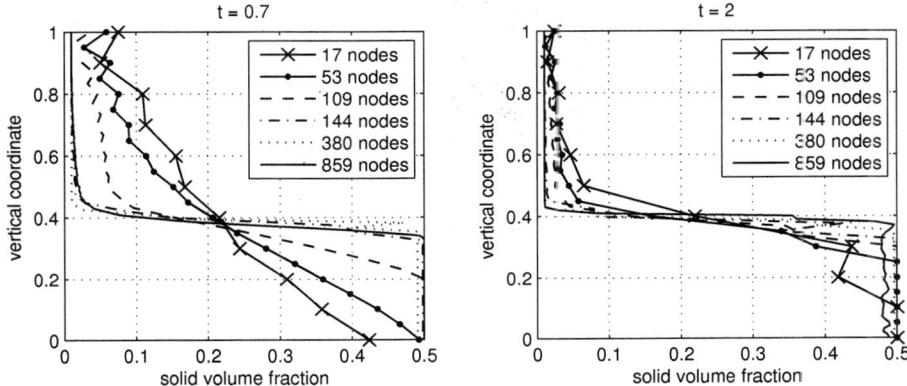

Figure 3.10: Vertical profiles of solid volume fraction at $t = 0.7$ and $t = 2$. Formulation with $p\nabla C_s$-term.

3.7.3 Flow over a backward-facing step with free surface

In this analysis we look at the free-surface flow of a two-phase mixture over a backward-facing step. The flow is driven by the velocity prescribed at the inflow boundary $\mathbf{v}_s = \mathbf{v}_f = [5\ 0]^T$, and a body force acting at an angle: $\mathbf{b} = [1\ -10]^T$. On the other fixed boundaries the velocities are set to zero. No pressure Dirichlet condition is applied. The mixture entering at the inflow boundary has a constant solid volume fraction of $C_s = 0.6$. The maximum solid volume fraction is set to $C_s^{max} = 0.8$. The same definition of the drag force as in the sedimentation test (Section 3.7.2) with $F' = 10$ is used. The parameters of the model are summarized in Table 3.6 and a typical mesh with the model dimensions is shown in Figure 3.11.

phase	ρ	μ	F'	Δt	N_{nodes}
solid	2000	1000	10	0.02	275-424
fluid	1000	1			

Table 3.6: Parameters used in two-phase flow over backward-facing step.

In Figure 3.12 the solid phase velocity vectors are plotted. The fluid is colored according to the solid volume fraction. The results show a very smooth free surface. The volume fraction of the solid phase tends to increase at the bottom of the flow. Behind the step the fluid phase accumulates. This can be attributed to the low pressure, caused by the sudden enlargement. While the high-viscosity solid phase flows past the sharp corner the fluid phase gets sucked

90 *Chapter 3* – **Multi-phase free-surface flows**

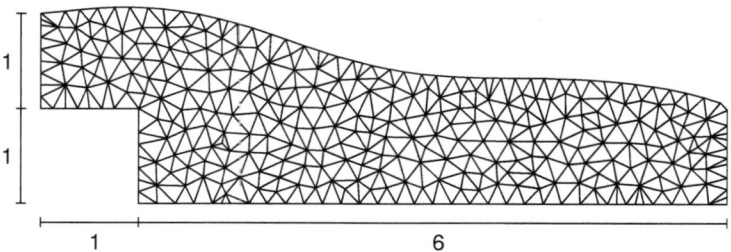

Figure 3.11: Two-phase flow over a backward-facing step: Unstructured mesh at $t = 20$.

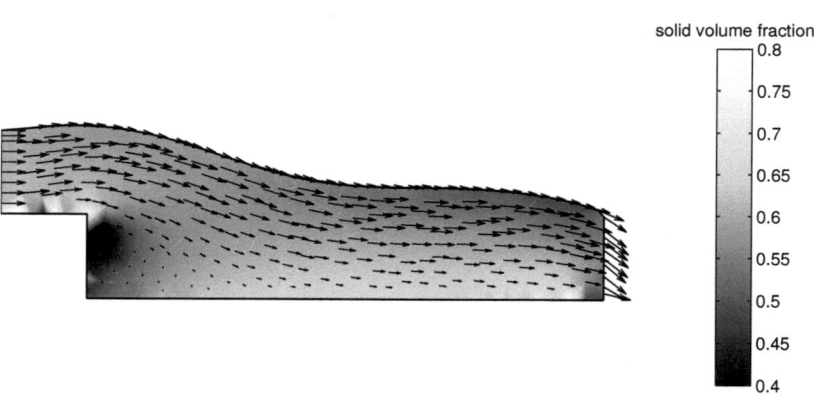

Figure 3.12: Two-phase flow over a backward-facing step: Solid volume fractions and velocity vectors of the solid phase are shown at $t = 20$.

3.7 – Numerical tests

into that zone. The volume fractions are smooth throughout the domain, with the exception of some nodes on the fixed boundary.

3.7.4 Sharp gradients of volume fractions

Next we look at how well two initially separated phases remain separated in the case of a very high drag force coefficient K_{drag}, acting as a penalty on the velocity difference between the two phases. The velocity difference between both fluids is thereby forced to be zero throughout the domain. The deviation of volume fractions from their initial values $C_s^1 = 0.01$ and $C_s^2 = 0.99$ is a measure for the performance of the numerical method.

In the results shown subsequently both phases are present throughout the domain. This means that in areas filled with mainly one phase the other phase is still present with a volume fraction close but not equal to zero. The volume fraction of one phase cannot become zero for numerical reasons, the stiffness matrix would become ill defined.

3.7.4.1 Vertical separation

A dense and a lighter fluid are separated by a vertical line running through the center of a rectangular container. The dense fluid eventually displaces the less dense fluid at the bottom of the container and the volume fraction gradients are oriented in vertical direction. The computation is performed using a fixed mesh onto which all variables are mapped after the nodes have been moved at the end of each time step. The two-phase fluid is characterized by the following properties: Densities $\rho_1 = 2000$ and $\rho_2 = 1000$, viscosity $\mu_1 = \mu_2 = 100$ and the drag coefficient $K_{drag} = 10^7$. The model consisting of about 2900 nodes was analyzed in 500 time steps of length $\Delta t = 0.02$ each. Figure 3.13 shows contour lines of constant volume fraction of the denser phase. We can clearly see the diffusion of the initially sharp gradient over about 4 elements during the first few time steps. Later in the analysis this diffusion appears to stabilize as the overall motion in the container slows down. This diffusion is mainly artificial and is caused by the smoothing effect of the method employed to compute volume fractions. However its extent is not dramatic.

Another undesired effect manifests itself where the phase separation touches the boundary. Since in this problem we used no-slip boundary conditions on all boundaries the volume fractions at boundary nodes cannot migrate by convection, but only by the movement of adjacent interior nodes. Thus the phase that is being moved tends to 'stick' to the boundary, especially near corners. However this effect is contained within a thin boundary layer whose thickness can be reduced by mesh refinement.

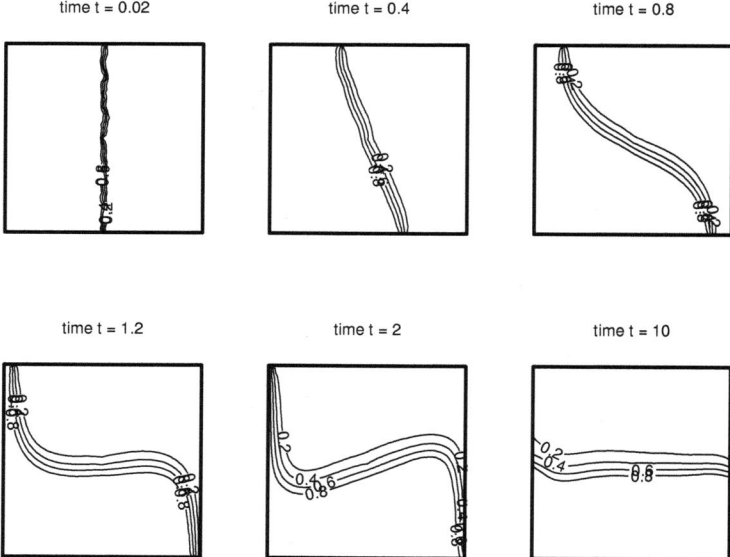

Figure 3.13: Contour lines of constant volume fraction of the denser phase for two initially vertically separated fluids.

3.7.4.2 Sinking drop

A circular drop of a heavy fluid is released inside a lighter fluid. The simulation is carried out on a square domain with a free surface at the top. The following material properties were used in this analysis: densities $\rho_1 = 2000$ and $\rho_2 = 1000$, viscosities $\mu_1 = \mu_2 = 1$ and the drag coefficient $K_{drag} = 10^5$. The fluid is discretized using roughly 3000 nodes and 200 time steps of $0.01s$ were computed. In Figure 3.14 contours of volume fractions of the denser phase are displayed. Here

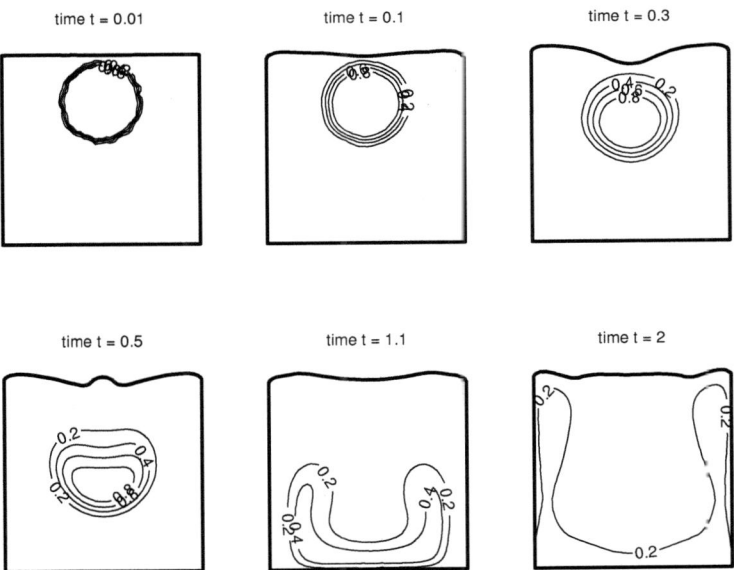

Figure 3.14: Contour lines of constant solid volume fraction for sinking drop of heavy fluid.

we also observe diffusion of the gradient of volume fraction at the separation between phases, an effect which is mostly due to the method for computing volume fractions. Compared to the previous simulation the diffusion of the gradient continues until, in the end, the maximum volume fraction of the denser phase has decreased to about 0.2. This simulation points out a weakness of the present numerical method for problems where strong variations of volume fractions have to be represented. We can however point out that this numerical diffusion can be reduced by refining the mesh in regions of high volume fraction gradients.

3.7.5 Dam break

The dam-break test problem with a two-phase material represents one further step towards simulating a mudflow event. On the same simple geometry as in the single-phase test we analyze the behavior of a two-phase mixture as it moves forward, pushing the free surface along a free-slip boundary. Two simulations using different drag force coefficients K_{drag} illustrate the ability of the method to follow the flow of a mixture of two materials with varying volume fractions. The material properties used in these simulations are: Densities $\rho_s = 1000$ and $\rho_f = 500$, viscosities $\mu_s = 100$ and $\mu_f = 0.1$ and $K_{drag} = 100$ in the first simulation and $K_{drag} = 10'000$ in the second. 150 time steps of length $\Delta t = 0.005$ are computed using a mesh of approximately 650 nodes. In Figure 3.15 we show solid volume fractions on the deformed fluid masses at different time steps. While the solid phase rapidly accumulates at the bottom due to gravity

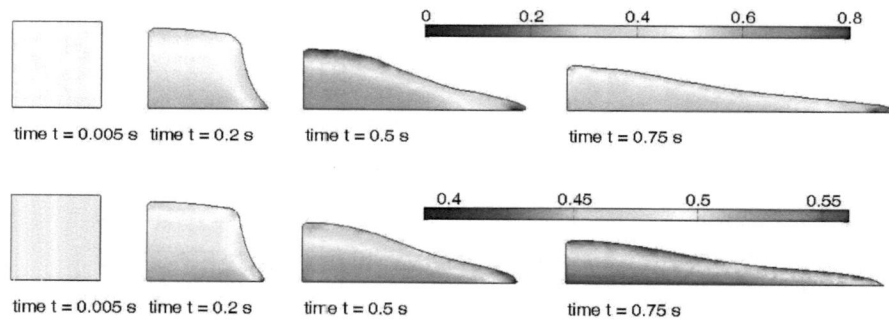

Figure 3.15: Dam-break simulation: Solid volume fractions are shown for two different drag coefficients. Top: $K_{drag} = 100$, bottom: $K_{drag} = 10'000$.

the fluid accumulates at the front and at the top free surface. We note that the variation of the volume fraction is much larger in the case of a low drag force coefficient. However the shape of the flowing mass is very similar in both cases.

3.7.6 Mudflow impacting an obstacle

After having verified all the major components of the mudflow model it is time to see if the method is capable of solving the original problem: The simulation of the downhill propagation of a two-phase mixture and its impact on an obstacle. The obstacle, representing a protection

3.7 – Numerical tests

dam, is modeled as a solid block which is placed at the bottom of a slope. The geometry, together with the mesh at time $t = 0$ is given in Figure 3.15. The flow is initiated by the sudden

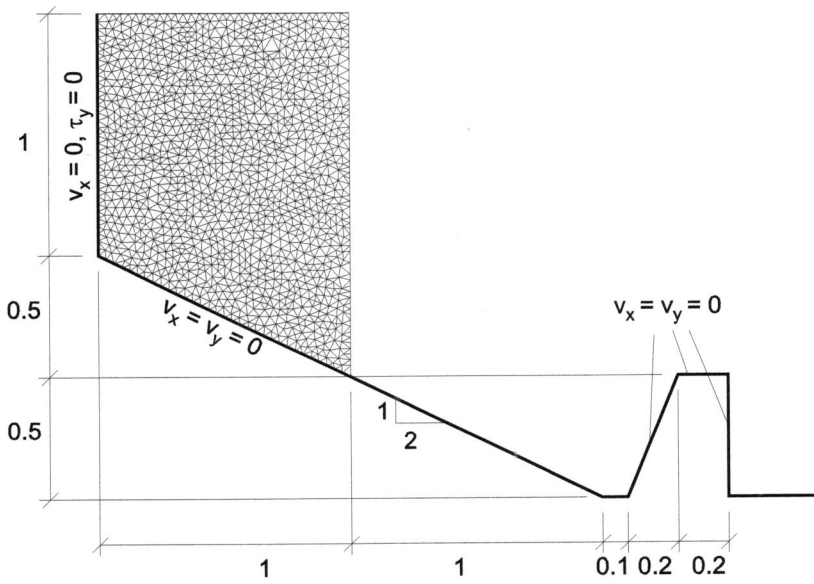

Figure 3.16: Geometry and initial mesh used in the simulation of a mudflow impacting on an obstacle.

release of a homogeneous two-phase mixture. Material parameters together with details of the discretization are given in Table 3.7. For comparison the same problem is simulated with a single-phase fluid, using the average material properties of the two-phase mixture.

	ρ_s	ρ_f	μ_s	μ_f	K_{drag}	C_s^{init}	Δt	No. of nodes
Two-phase model	1000	500	100	2	10'000	0.5	0.002	≈ 1550
Single-phase model	750		51		-	-	0.002	≈ 1550

Table 3.7: Parameters of the simulation of a mudflow impacting on an obstacle.

The shape of the two-phase fluid during the event is shown in Figure 3.17. Colors indicate solid volume fractions. The solid phase initially accumulates at the base, while the fluid phase

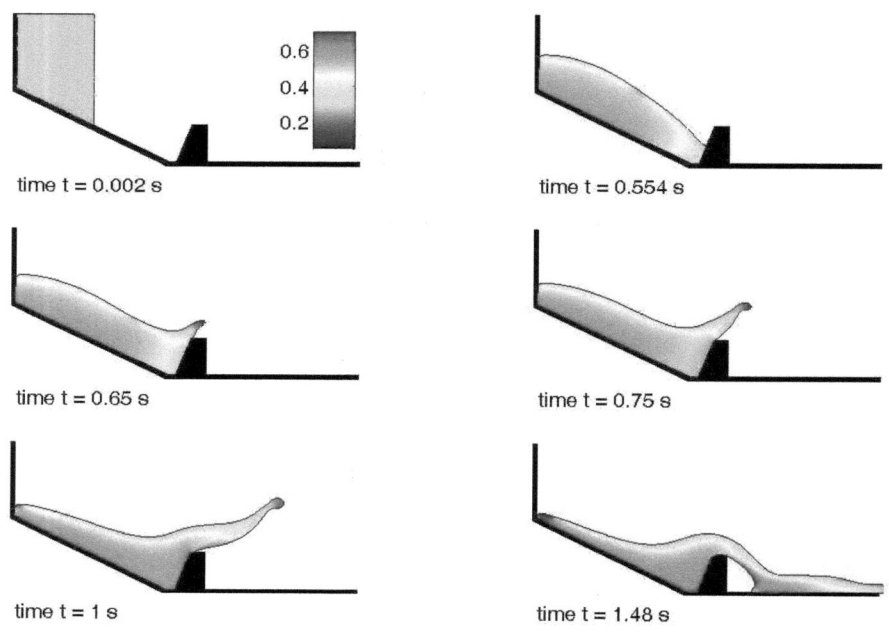

Figure 3.17: Solid volume fractions on a mudflow impacting on obstacle.

stays on the surface and accumulates at the front of the flow. After the flow tip reaches the obstacle the solid phase quickly catches up with the faster flowing fluid phase, filling the space behind the barrier. The tip of the mixture shooting over the barrier is essentially fluid, due to lower viscosity and density. At the end of the simulation, that is after 1000 time steps, the total volume of the mixture has increased by 2.3%. We consider this error very small, considering the relatively simple contact algorithm used in the model.

From the simulation the resultant force acting on an obstacle that obstructs the flow path is extracted. In Figure 3.18 the resultant force of the single-phase and the two-phase model are compared. The force is computed by integrating the pressure along the front side of the obstacle. Right after the impact the force attains its peak, before it slowly decays to the hydrostatic

3.7 – Numerical tests

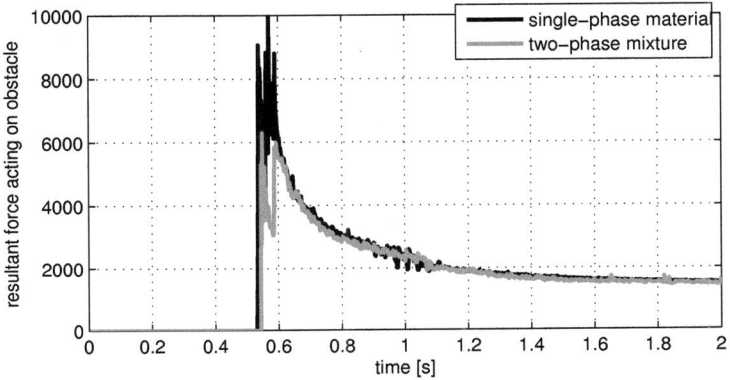

Figure 3.18: Resultant force acting on obstacle.

level. The peak force right after the impact is higher in the case of a single-phase material. This can be explained with the lower density of the fluid phase, which reaches the obstacle first in the two-phase simulation. After a while however the difference vanishes. The fluid phase acts as a buffer, attenuating the impact of the denser solid phase.

In an attempt to estimate the momentum delivered to the obstacle we consider conservation of momentum of the part of the fluid that remains behind the obstacle. The balance equation is written along an axis perpendicular to the obstacle.

$$F = F_{hs} + F_{dyn} \qquad (3.7.6)$$

where F is the force acting perpendicular to the wall (plotted in Figure 3.18), F_{hs} the hydrostatic pressure integrated over the face of the obstacle and F_{dyn} the dynamic force due to the slowing down of the fluid mixture. We assume that the force F reaches its permanent value F_{hs} at $t_1 = 1.4s$. All computations are performed for a slice of $1m$ depth in the third dimension.

Assuming that the average of the solid volume fractions is close to 0.5 the hydrostatic pressure acting on the obstacle can be evaluated as

$$F_{hs} = \frac{p_1 + p_2}{2} d = \frac{1500 + 5250}{2}(0.2^2 + 0.5^2)^{0.5} \approx 1800 kN \qquad (3.7.7)$$

The change of momentum of the fluid between $t_0 = 0.55s$ and $t = 1.4s$, $I = \int_{t_0}^{t_1} F_{dyn}\, dt$, can be obtained by multiplying the change of velocity by the mass of the fluid, that remains behind the obstacle:

$$I = \Delta v \rho V \qquad (3.7.8)$$

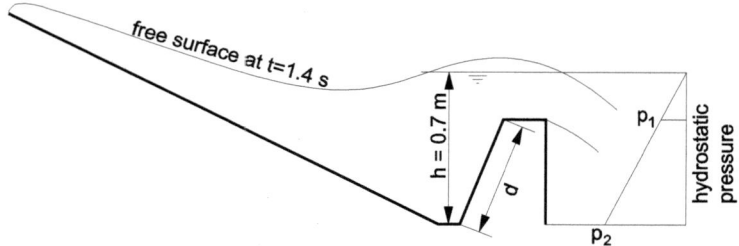

Figure 3.19: Illustration of the problem geometry with free surface at $t = 1.4s$.

Δv is the average velocity before the fluid reaches the obstacle. Assuming a parabolic distribution in the direction perpendicular to the slope we can estimate that $\Delta v \approx 2.3 m/s$. With the volume $V = 0.5 m^3$ and $\rho = 750 kg/m^3$ the change of momentum is $I \approx 860 kNs$.

On the other hand, the momentum transferred to the obstacle is obtained by computing the surface below the curve in Figure 3.18. For the two-phase mixture, and after subtracting the hydrostatic force, this evaluates to

$$\int_{t_0}^{t_1} = (F - F_{hs})\, dt \approx 760 kNs \tag{3.7.9}$$

The result of the hand calculation overpredicts this value by about 13%. This represents an additional verification of the numerical method.

In the hand calculation several approximations have been made. The average velocity Δv and the hydrostatic force are difficult to estimate. However, it is possible for mudflow events to obtain velocities that only depend on the material properties and on the slope angle. This can be done by performing parametric studies varying the angle of an infinite slope. Thus we can imagine to use the present two-phase model to obtain rules of thumb for estimating loads acting on protection structures.

3.8 Conclusions

In this chapter we developed a framework for modeling two-phase flows that undergo large motions. The governing equations of the two phases have been derived from the equations of a single-phase Newtonian fluid by applying mixture theory. By smoothing both phases over the computational domain we obtain a formulation where the presence of a material is given by a volume fraction in each node. The method is implemented in a finite element framework, where

3.8 – Conclusions

each node has 5 degrees of freedom (in two dimensions) The velocities of both phases and a pressure, which is common for both phases. The Lagrangian update of a node is performed for both phases, resulting in two updated positions for each node. A re-meshing step creates a new mesh of good quality, on which the nodal values are interpolated from the previously updated nodal positions.

The correct implementation of the two-phase Newtonian fluid is verified by comparing it to the single-phase solution. Being able to match the single-phase behavior as a limit case is an important result. It shows that the algorithms and methods inherited from the single-phase model, the stabilization most importantly, but also the time stepping algorithm and the mesh update procedure remain fully valid for the two-phase formulation.

The method for computing volume fractions at the end of each iteration yields smooth and accurate results. For sharp gradients of volume fractions test problems pointed out to which extent numerical diffusion of the gradients occurs. A set of tests, where the method has been de-coupled from the physics of two-phase flow, has, however, also shown that mesh refinement is capable of effectively remediating this problem.

The sedimentation test allowed to show that the method can easily be adapted to match the results of an analytical solution without any parameter calibration. Being able to match the analytical result using a particular constitutive model on a simple test set-up opens the door for the implementation of more complex constitutive models.

Further numerical tests investigated the behavior of the method with free surfaces and showed that the method very accurately conserves the mass of the mixture. Finally we demonstrated that problems representing geophysical flows can be analyzed using the numerical method. We conclude that the proposed algorithmic framework is capable of following the motion of two-phase mixtures in a wide range of problem types.

Chapter 4

Implementation

4.1 Introduction

4.1.1 Object-oriented finite element programming

Object-oriented programming provides a way to organize a computer code, grouping subroutines and variables into classes in order to facilitate the maintenance of large projects and make it easier to extend the code with new functionality. Object-oriented programming helps a developer to organize in a structured way the tasks he wants the program to perform. Finite element methods are particularly well suited for being implemented in an object-oriented framework. The modular aspect with a mesh, which is composed of elements and nodes, as well as the ease of implementing slightly modified formulations by making use of the concept of inheritance have contributed to the success of object-oriented programming in the field of finite element modeling. In the present work the implementation is based on FEM_object, a finite element program for static and dynamic nonlinear analysis in solid mechanics [1]. An in-depth discussion of concepts of object-oriented programming together with a detailed description of the code is available in Commend et al. [13].

4.1.2 An object-oriented framework for finite element modeling on moving domains

The global organization of the code corresponds to a semidiscrete implementation, using a time stepping algorithm to advance the solution in time. At each time step we solve a nonlinear boundary-value problem iteratively. In the particular case of the two-phase updated Lagrangian description, the mesh is regenerated after each iteration. The key object of the implementation is the Domain, which sends messages to other classes to build a linear system of equations and solve it at the current time step. The flow chart in Figure 4.1 illustrates the interaction between the main objects in the solution procedure for a general problem solved on a moving domain.

The solution procedure starts with the creation of a Domain. The Domain reads the problem definition from an input file. On one problem definition several types of analyses can be performed, for example an eigenvalue analysis, an initial state analysis or a safety factor analysis. We focus on a transient analysis. The main steps of a transient analysis are explained in the following. The numbers refer to a specific task in Figure 4.1.

1. The Domain creates an Analysis-object of type TransientAnalysis and sends a message to execute the analysis.

---
[1]http://www.zace.com/femobj_nl/femobj_nl.htm

4.1 – Introduction

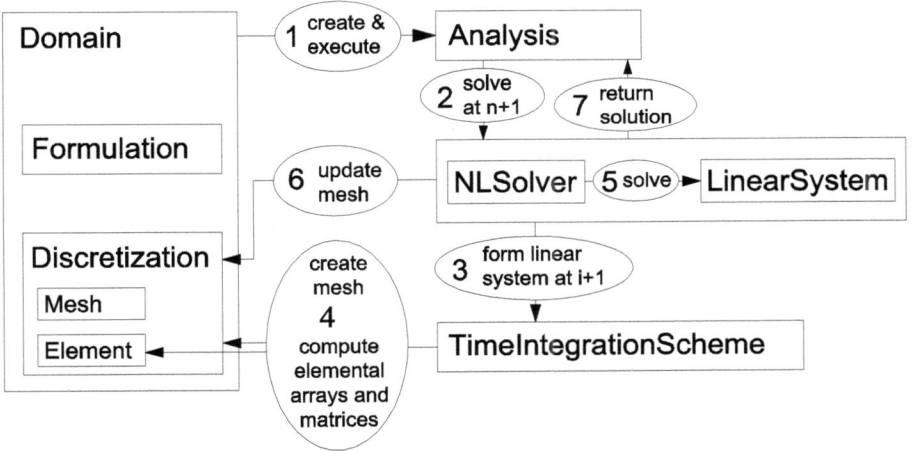

Figure 4.1: Simplified illustration of interaction between classes of object-oriented computer program.

2. The `TransientAnalysis` then sends a message to the `NLSolver` to solve the problem at time t_{n+1}.

3. The `NLSolver` creates a `LinearSystem` and send a message to a `TimeIntegrationScheme` to compute the right-hand-side and the left-hand-side of the `LinearSystem` at iteration $i+1$.

4. The `TimeIntegrationScheme` loops over all elements in order to assemble the `LinearSystem`. In order to do this a message is sent to the `Discretization` to create a mesh containing `Elements` and `Nodes`. The specific type of `Elements` and the form of the elemental matrices and arrays is given by the `Formulation`.

5. After the `LinearSystem` has been assembled it is solved and the solution vector is returned to the `NLSolver`.

6. The `NLSolver` sends a message to the `Discretization` to update the mesh according to the solution vector.

7. If the convergence criteria of the `NLSolver` have been met the solution is returned to the `TransientAnalysis`, where the results for the time step t_{n+1} are written to the output files.

4.2 Organization of the research code

The development of the code has been guided by the requirement that all problems in this work are to be solved by the same computer program. The input data is read from a text file, which specifies all the aspects of the analysis. This approach simplifies version management of the code and verification of all the parts of the implementation is improved since the same methods are used for a wide variety of problems.

New functionality was built into the code by using existing classes as much as possible. Some parts of the structure of the original FEM_object however had to be substantially modified in order to accommodate for instance moving meshes, unstructured meshes, meshless methods and different formulations. The main points that required modification are listed below.

Moving meshes: Meshes that evolve with time require a lot of geometry-related tasks to be performed on the elements and nodes. This led to the separation of numerical model-related data, contained in the nodes and elements, from geometry-related data, contained in points and triangles. A new class `Mesh` was introduced that stores the geometric information of a mesh at a specific iteration (see Section 4.2.3). Operations such as applying a

4.2 – Organization of the research code

displacement field or re-creating a new triangulation can be performed on a Mesh-object. Mapping of field variables can be performed between two Mesh-objects. The possibility to include both fixed (Eulerian) and moving (Lagrangian) meshes in the same code prompted the creation of the Discretization-class.

Formulations: In this work we implemented a two-phase formulation where both phases share the same pressure. In order for the code to be able to accommodate other formulations, for instance with an additional solid, grain-to-grain pressure, a new structure defining the field variables had to be created. The idea, presented in Eyheramendy [20], consists in creating a class Formulation, in which the definition of the Element specific to that formulation is given as an embedded inner class. The combination of the classes Formulation and Element define the complete behavior of the underlying governing equations. In order to re-use the same elemental arrays and matrices for several different formulations, a new class ElementMatrices has been created. The hierarchy of classes involved in the computation of elemental matrices and arrays is given in detail below.

Automatic meshing: In the original FEM_object, every Element and Node has to be specified as a line in the input file. This equivalence had to be abandoned due to automatic mesh generation. Now only a prototype of the Element or Node is specified in the input file, the nodal coordinates as well as element connectivities are provided by the mesh generator.

Use of third-party libraries: While the original FEM_object code is completely autonomous, the present research code makes heavy use of third-party libraries. The availability of standardized libraries for linear algebra, computational geometry and data types makes the FEM_object-classes Dictionary, List, FloatArray, IntArray and Matrix as well as their subclasses obsolete. The class Dictionary is replaced by the map-structure of the Standard Template Library (STL) in C++, while List is replaced by STL's list. For all linear algebra tasks the BLAS implementation in the boost::numeric library[2] is used. The sparse linear solver PARDISO[3], included in the Intel MKL library[4], increases speed while reducing memory requirements substantially compared to the skyline solver implemented in the original code. Finally, for all geometry-related tasks, the CGAL-package [9] is used. In order to access the library's functions an interface class CGALInterface is created.

---

[2] www.boost.org
[3] www.pardiso-project.org
[4] www.intel.com

The new class hierarchy is shown in Figure 4.2. The main changes with respect to the original FEM_object code are briefly discussed in the following.

4.2.1 Analysis class

The introduction of such a class is motivated by the separation of model data from the analysis. Such a subdivision is advocated in McKenna [37]. It provides a base class for different types of analyses, such as transient, static, eigenvalue, stability or others. The most important member function of the Analysis-class is solveYourself(), a method previously contained in the Domain.

4.2.2 Discretization class

The Discretization-class is introduced in order to accommodate different types of meshes. It contains a lot of the data that was previously in the Domain-class, in particular all the lists holding pointers to FEMComponent-objects. Two direct subclasses of the class Discretization are available: EulerianDiscretization and LagrangianDiscretization. While the coordinates of elements, nodes and Gauss points don't change in the EulerianDiscretization they need to be updated after each iteration in the LagrangianDiscretization. The most important members of this class are Mesh-objects, which contain the geometric information corresponding to a specific state of deformation. The Discretization-class has member functions which operate on Mesh-objects:

- giveNode(int), giveElement(int), giveBoundarySegment(int) etc.: Return existing FEMComponent or creates new one.

- printOutput(): Prints results to file(s).

- initializeMeshForStep() and terminateIteration(): Group all the operations that have to be performed in order to update the Mesh according to the new predictor or corrector displacement increment. These operations are:

    - updateNodes() Changes the coordinates of the nodes according to the displacement increment given by the Lagrangian update
    - updateMesh() Updates the Mesh-object. Depending on whether re-triangulation and/or re-zoning is needed, moves the existing GeometryComponents or re-creates them.

4.2 – Organization of the research code

CGALInterface
Discretization
    EulerianDiscretization
    LagrangianDiscretization
        LagrangianTwoPhaseDiscretization
Dof
Domain
ElementMatrices
    TwoPhaseElementMatrices
FEMComponent
    Analysis
        EigenValueAnalysis
        InitialStateAnalysis
        StaticAnalysis
        TransientAnalysis
    BCLine
    BoundarySegment
    Element
        MixedUP
        DisplacementBased
        TwoPhase
    Formulation
        MixedUPFormulation
        DisplacementBasedFormulation
        TwoPhaseFormulation
    Load
        BodyLoad
            DeadWeight
        BoundaryCondition
        NodalLoad
    LoadTimeFunction
        ConstantFunction
        PiecewiseLinFunction
    Material
        ElasticMaterial
        VonMisesMaterial
        NewtonianFluid
        TwoPhaseFluid
    NLSolver
    Node
    ShapeFunction
        ShapeFnFEM
        ShapeFnSibson
    TimeIntegrationScheme
        Static
        Trapezoidal
        Newmark
    TimeStep
Stabilization
    PressureGradientStabilization
        ConsistentStabilization

Continued on next page

Fields
GaussPoint
GeometryComponent
    Line
    Point
    Triangle
Mesh

Figure 4.2: Class hierarchy. Additions to the original FEM_object are printed in **bold** letters. Classes that have not been implemented are printed in *italics*.

- `updateFEMComponents()` Updates or creates `FEMComponents` based on the updated `Mesh`.
- `remapFields()` Maps field variables from the old to the new `Mesh`.

4.2.3 Mesh class

Most of the problems in numerical modeling using Lagrangian meshes are related to the temporally varying geometry. In order to be able to efficiently update the mesh and map variables from one mesh to another we need a robust way to store all geometric data such as nodal positions, element connectivities, fixed and free boundaries and so on. Since most of these data structures as well as operations performed on them are standard geometrical problems we chose to make use of the Computational Geometry Algorithms Library (CGAL) [9].

The geometrical information related to one mesh at a given time is stored in a `Mesh`-object. This object contains only information relevant to the geometry and is completely dissociated from the physics of the problem. It contains no information such as nodal velocities, material properties or elemental stiffness matrices. `Mesh`-objects constitute the interface between the CGAL-library and the finite element code. In the following we distinguish between the geometrical objects of type `GeometryComponent`, namely points, triangles and line segments, which are contained in the `Mesh`, and the corresponding finite element objects: Nodes, elements and boundary edges.

The `Mesh` is used to create the objects of the finite element method (nodes, elements and boundary edges). The process of mesh movement, re-meshing and re-mapping for the two-phase formulation is illustrated in Figure 4.3. At the end of each iteration the coordinates of all points are updated according to the computed displacement increments while all connectivities are retained. This update yields two deformed `Mesh` objects, one for each phase. A new `Mesh` is then created based on the α-shape of all the nodes of the deformed `Meshes` of the two phases. More specifically, the nodes of both deformed `Meshes` are inserted into a new Delaunay tri-

4.2 – Organization of the research code

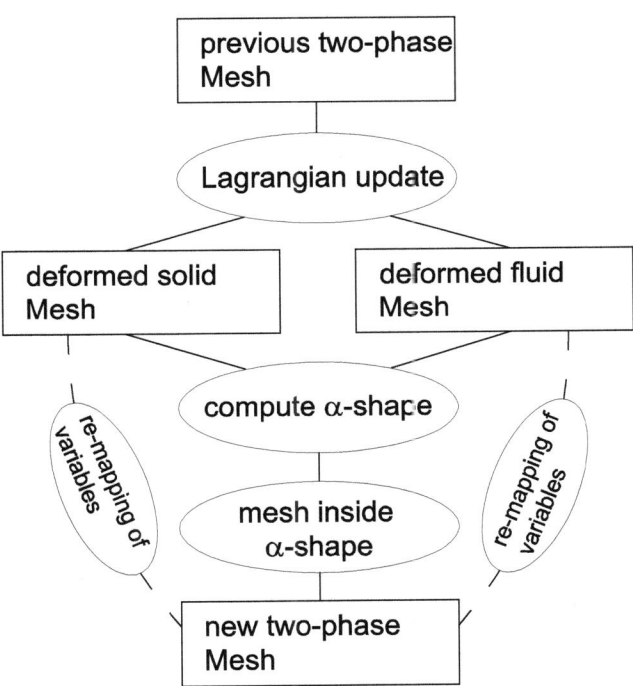

Figure 4.3: Flow chart illustrating the process of mesh movement, re-meshing and re-mapping.

angulation and the boundary of the α-shape, computed according to Appendix A.2, is stored. Subsequently the boundary segments of this new Mesh are inserted as constraints into a constrained Delaunay triangulation and new nodes are inserted such that the new triangular mesh satisfies the mesh quality requirements. Finally, the variables of the finite element method are mapped from the two deformed Meshes onto the new Mesh, following the procedure outlined in Section 2.5.5.

4.2.4 Computation of elemental matrices and arrays

The computation of elemental arrays and matrices is re-organized substantially with respect to the original code. A new class ShapeFunction, whose main task it is to compute shape functions and return their values to a Gauss point, is created. Two subclasses exist, FEMShapeFn and SibsonShapeFn. The hierarchy related to the computation of elemental arrays and matrices is illustrated in Figure 4.4.

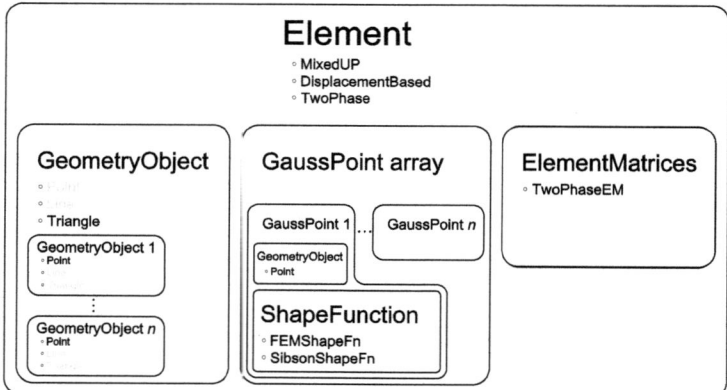

Figure 4.4: Hierarchy of classes related to the class Element. Subclasses are indicated as bullet points below the superclass. Subclasses printed in grey are not applicable in the given configuration.

The GeometryObject itself is not stored in the Element, only a pointer to the corresponding object in the Mesh-class.

4.2.5 Assembly of RHS and LHS

In the original FEM_object code the RHS and LHS (Right- and LeftHandSide) are assembled in the `Domain`-object. We think that this task is better accomplished in the class `TimeIntegrationScheme`. This way the same member functions of the `Element`-class can be called for all time integration schemes, while no information about the scheme is required in the element. The method for computing the left hand side for the generalized trapezoidal algorithm is given in Figure 4.5.

```
void    Trapezoidal :: ComputeLeftHandSide(Solver * linearSolver)
{
   Element          *element;
   double_matrix    C,M;
   int_vector       loc;

   Discretization   *aDiscretization;
   aDiscretization = this -> giveDomain() -> giveDiscretization();

   double dt        = this -> giveCurrentStep() -> giveTimeIncrement();
   double gamma     = this -> giveGamma();
   double factor    = 1./(dt*gamma);

   int nElements = aDiscretization -> giveNumberOfElements();
   for (int i=1; i<=nElements; i++){
      element   = aDiscretization -> giveElement(i);

      C   = element -> giveViscosityMatrix();
      M   = element -> giveMassMatrix();
      C  += factor * M;

      loc = element -> giveLocationArray();

      linearSolver -> assembleLHS(C, loc);
   }
}
```

Figure 4.5: Method `ComputeLeftHandSide` as a member function of the class `Trapezoidal`.

4.2.6 Other additions

Other classes that have been added to the original FEM_object are briefly summarized below.

BCLine: This class is required in order to impose boundary conditions on a moving domain. After each Lagrangian update of the nodes we test if the node at its new position is in contact with the boundary (see Section 2.5.6). The position of the fixed boundary is specified by straight line segments, defined by objects of the type BCLine.

BoundarySegment: In order to apply surface loads, such as surface tension, the boundary of the computational domain has to be specified. The element edges, which form the exterior boundary, are identified during re-meshing. Pointers to these BoundarySegments are stocked in a list which is a member of the Discretization.

Stabilization: The creation of a Stabilization-class facilitates the implementation of different variants of stabilization methods. Each Element has a member of type Stabilization, whose main member functions are ComputeStabilizationMatrix() and ComputeStabilizationForceVector. In order to compute these matrices and arrays the Stabilization makes use of the ElementMatrices of the Element.

4.3 Spatial searching

Efficient spatial search algorithms have been crucial for the development of the computer code used in this work. Since the spatial coordinates of nodes change continuously the information on spatial proximity has to be redefined after each Lagrangian update. Proximity information is needed for the following operations:

- Re-mapping of variables from one mesh to another by linear interpolation requires for each point of the new mesh the containing triangle of the previous mesh to be found.

- After re-creating a new mesh within the boundary of the previous mesh using a conforming Delaunay mesher the triangulation covers the entire convex hull of the set of nodes. The detection whether a triangle lies inside or outside the domain boundary results in a point-in-polygon problem, a well-known problem in computer graphics. We chose the ray-casting algorithm, described for example in Sutherland et al. [51]. This algorithm consists in computing the number of times that a ray, starting from the center of a triangle and going to infinity, intersects with the boundary of the polygon. If the number of intersections is odd, then the triangle is outside the computational domain, if it's even, then the triangle is inside the domain. This search for intersections can be greatly accelerated if we limit the search to boundary segments that lie inside a slice of finite size surrounding the ray.

- Contact detection of nodes with the fixed boundary involves the search of nodes that are located within a certain distance of a boundary segment. This search can become expensive if the fixed boundary is discretized with a large number of segments.

The spatial search algorithm by Munjiza et al. [38] is very simple and easy to implement. The algorithm divides the domain into an evenly spaced grid ordered along the coordinate axes. In a first step, a loop over all container objects, for example triangles for the point-in-triangle search, is performed and a list of all objects that touch a grid cell is created. Once this structure is generated the search for a container object of a point can be narrowed down to all container objects that touch the grid cell in which the point is located.

Time complexity of the algorithm is of order $O(n)$ with respect to the number of points to be located. Memory requirement is dependent on the size of the grid cells. Therefore the best choice of grid cell size is a trade off between memory requirements and speedup. For the point-in-triangle search a grid cell size that leads to an average of about 5 triangles per cell has been found to be optimal.

4.4 CPU-time

Figure 4.6 shows CPU-times versus number of nodes for a series of computations of the two-phase dam break problem (see Section 3.7.5). In the range up to about 2000 nodes CPU-time scales about linearly, whereas above 2000 nodes we notice a sudden increase of the slope in the *log-log*-plot. This is due to the size of the cache memory. Above a certain number of triangles the spatial search data structure becomes too large to fit into the fast memory of the CPU and memory access slows down the total time of the computation.

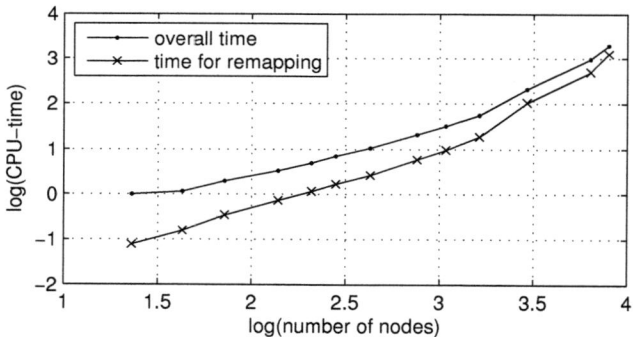

Figure 4.6: CPU-times as a function of nodes for the two-phase dam break problem.

Chapter 5

Concluding remarks

5.1 Conclusions

Specific conclusions regarding the single-phase formulation can be found in Section 2.7, and for the two-phase formulation in Section 3.8.

A new and innovative numerical method for simulating two-phase free-surface flows has been developed in this work. The method is capable of simulating a wide range of problem types from sedimentation of solid material in a fluid to gravity-driven free-surface flows. The key feature of the method is an algorithm that allows the motion of two different constituents of a mixture to be followed in a Lagrangian reference frame.

In contrast to existing debris- or mudflow models the method implements a continuum approach, which allows to obtain detailed time histories and profiles of stresses, velocities or volume fractions. The algorithmic framework is kept general in order to allow any kind of constitutive model to be included. We expect the method to find a wide range of applications not only in the field of geophysical flows, but in any kind of problem involving the motion of two phases where interaction between the phases cannot be neglected.

In the following we outline some possible directions of future research that are expected to either improve the performance of the current method or extend its range of applications.

5.2 Further research

5.2.1 Constitutive modeling

The first step towards a more realistic debris- or mudflow model has to go in the direction of constitutive modeling. Particularly assumptions B and C in Table 3.1 have to be investigated more closely. In more heterogeneous mixtures with a large fraction of coarse-grained material grain-to-grain contact is playing an important role. We suggest further development to be directed towards the implementation of models similar to the one presented by Hutter et al. in [30].

5.2.2 Pseudo three-phase formulation

In this work we assumed the two-phase fluid to be fully saturated at all times (Assumption D in Table 3.1). This assumption is probably reasonable for mixtures of uniform grain-size distribution. When the volume fraction of relatively coarse-grained material such as gravel and rocks is high, then some parts of the mixture are likely to become partially saturated. Especially at the free surface the fluid can seep through the granular material, and air can fill up the inter-

5.2 – Further research

granular voids. In this case Equation 3.1.2 has to be modified in order to consider the volume fraction occupied by air, C_a.

$$C_s + C_f + C_a = 1 \qquad (5.2.1)$$

The solid volume fraction C_s cannot exceed a maximum value which is given by the inter granular voids. For gravel this value is typically around $C_s^{max} = 0.6 - 0.8$.

We assume that the air can take the place of the fluid, but not the solid. The air and the fluid are assumed to move with the same velocity, thus the term 'Pseudo three-phase formulation'.

Taking the derivative with respect to time of Equation 5.2.1 on a fixed reference frame gives

$$\frac{\partial C_s}{\partial t} + \frac{\partial C_f + C_a}{\partial t} = 0 \qquad (5.2.2)$$

This leads to a modified equation of conservation of the volume of the mixture:

$$\nabla \cdot (C_s \mathbf{v}_s) + \nabla \cdot ((C_f + C_a)\mathbf{v}_f) = 0 \qquad (5.2.3)$$

At the boundary the Lagrangian update has to consider the new mixture velocity

$$\mathbf{v}_m = C_s \mathbf{v}_s + (C_f + C_a) \mathbf{v}_f \qquad (5.2.4)$$

The material properties of the fluid phase also need to be modified:

$$\rho'_f = C_f \rho_f + C_a \rho_a \qquad (5.2.5)$$
$$\mu'_f = C_f \mu_f + C_a \mu_a \qquad (5.2.6)$$

ρ'_f and μ'_f are the averaged fluid phase properties used in the computation of the body force, the mass and the stiffness matrices. ρ_a and μ_a can be assumed to be zero.

The computation of volume fractions after each iteration would in this formulation remain essentially the same. The volume of air would be considered part of the volume of fluid. The correction of the volume fractions, point 3 in Table 3.4, would be replaced by the computation of the new volume fraction of air:

$$C_a = \max(0, 1 - C_s + C_\tau) \qquad (5.2.7)$$

5.2.3 Fluid-structure interaction

As we mentioned in the introduction, the method developed in this work is expected to find its application in the design of retaining structures for mud- and debris flows. Section 3.7.6 gives an example where forces acting on a rigid structure have been computed. In order to go one step further, the interaction between flexible protection barriers and a flow mass has to be analyzed. The inclusion of fluid-structure interaction capabilities in the current formulation requires several extensions. The time stepping algorithm has to be adapted for the inclusion of stiffness terms, the structure undergoing large deformation needs to be modeled and an appropriate contact algorithm at the interface between the fluid and the structure has to be implemented.

5.2.4 Remarks for extension to three spatial dimensions

Although the method for updated Lagrangian simulation of two-phase free-surface flows presented herein is developed in two dimensions the extension to three dimensions is mostly straightforward. In question are the following components of the method: The governing equations, the finite element approximation, the creation of the mesh, the boundary contact detection and the re-mapping of variables. In the following we give some insight into the extension to 3D of each of these components, giving references for further reading, but without going into too much details.

Governing equations: The governing equations given previously remain fully valid in 3D.

Finite element approximation: Instead of using linear triangular finite elements with equal order interpolation for velocities and pressure linear tetrahedra can be used with the same stabilization method remaining valid.

Mesh creation: Algorithms for the construction of 3D Delaunay triangulations as well as 3D α-shapes are available in the CGAL package. Free tetrahedral mesh generators can be found on the internet.

Boundary contact detection: In order to allow the use of topographical data for the definition of the fixed boundary of the computational domain the boundary is best modeled as a triangular surface mesh. Contact detection between nodes of the fluid mixture and the boundary then consists of finding the position of a point with respect to polyhedral surface. For this search of proximity the same spatial search method as outlined in Section 4.3 using a three-dimensional grid can be used.

Re-mapping of variables: Linear interpolation in 3D also requires finding of the tetrahedra containing the point at which interpolated values have to be computed.

Appendix A

Appendix

A.1 Elemental matrices and arrays of the discrete weak form for single-phase flow

The elemental matrices for an element e that are assembled into Equation 2.5.19 are given by:

$$\mathbf{M}^e = \int_{\Omega_e} \rho \mathbf{N}^T \mathbf{N} d\Omega \tag{A.1.1}$$

The lumped mass matrix used in this work is obtained by summing the rows of the above matrix.

$$\mathbf{K}^e = \int_{\Omega_e} \mathbf{B}^T \mathbf{D} \mathbf{B} d\Omega \tag{A.1.2}$$

$$\mathbf{G}^e = \int_{\Omega_e} \mathbf{B}^T \begin{bmatrix} 1 \\ 1 \\ 0 \end{bmatrix} \mathbf{N}^p d\Omega \tag{A.1.3}$$

$$\mathbf{S}^e = -\tau_e \int_{\Omega_e} \nabla \mathbf{N}^{pT} \nabla \mathbf{N}^p d\Omega \tag{A.1.4}$$

$$\mathbf{f}^e = \int_{\Omega_e} \mathbf{N}^T \rho \mathbf{b} d\Omega + \int_{\Gamma_h} \mathbf{N}^T \mathbf{h} d\Gamma \tag{A.1.5}$$

$$\mathbf{f}^e_s = -\tau_e \int_{\Omega_e} \nabla \mathbf{N}^{pT} \rho \mathbf{b} d\Omega \tag{A.1.6}$$

The elemental matrix operators and elemental arrays are defined in terms of shape functions ϕ:

$$\mathbf{N} = \begin{bmatrix} \phi^1 & 0 & \cdots & \phi^n & 0 \\ 0 & \phi^1 & \cdots & 0 & \phi^n \end{bmatrix} \quad \mathbf{N}^p = \begin{bmatrix} \phi^1 & \cdots & \phi^n \end{bmatrix} \tag{A.1.7}$$

$$\mathbf{B} = \begin{bmatrix} \phi^1_{,x} & 0 & \cdots & \phi^n_{,x} & 0 \\ 0 & \phi^1_{,y} & \cdots & 0 & \phi^n_{,y} \\ \phi^1_{,y} & \phi^1_{,x} & \cdots & \phi^n_{,y} & \phi^n_{,x} \end{bmatrix} \quad \mathbf{D} = \begin{bmatrix} 2\mu & 0 & 0 \\ 0 & 2\mu & 0 \\ 0 & 0 & \mu \end{bmatrix} \tag{A.1.8}$$

$$\nabla \mathbf{N}^p = \begin{bmatrix} \phi^1_{,x} & 0 & \cdots & \phi^n_{,x} & 0 \\ 0 & \phi^1_{,y} & \cdots & 0 & \phi^n_{,x} \end{bmatrix} \tag{A.1.9}$$

$$\mathbf{b} = \begin{bmatrix} b_x & b_y \end{bmatrix}^T \quad \mathbf{h} = \begin{bmatrix} h_x & h_y \end{bmatrix}^T \tag{A.1.10}$$

The superscripts identify the nodes of the corresponding shape functions.

A.2 Delaunay triangulation and α-shapes

A triangulation of a set of points is the decomposition of the surface into triangles. The triangulation is called *Delaunay* if for each triangle, the circumscribed circle contains no points in

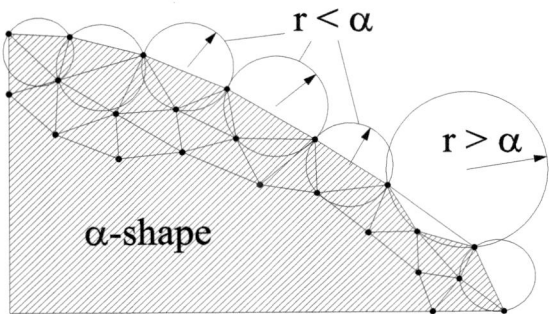

Figure A.1: Construction of α-shape

its interior. The concept of α-shapes is based on a Delaunay triangulation. It is widely used in computational geometry to transform a set of nodes into a shape consisting of surface and volume elements. The concept found its way into the computational mechanics community through Cueto et al. [14]. González et al. [21] use it for defining the free surface in updated Lagrangian fluid dynamics.

On the basis of the Delaunay triangulation only triangles (or tetrahedra in 3D) whose circumscribed circle (or sphere) have a radius not exceeding a user-defined value α are included in the α-shape. The basic idea is illustrated in Figure A.1. The computational domain is then set equal to the α-shape. The value α has to be chosen by the user in such a way that the important features on the free surface are most accurately represented by the α-shape.

A drawback of this method is the fact that the mass of a material modeled using α-shapes is not conserved. Lets imagine two adjacent finite element nodes on the free surface moving away from each other. At one point the circumcircle of the Delaunay triangle having both nodes as vertices will be too large and will therefore not belong to the α-shape anymore. Mass will be lost in this case. The opposite can occur as non-adjacent nodes on a concave surface move closer together, thus resulting in an increase of mass. These fluctuations in mass throughout the analysis can however be reduced by refining the nodal spacing.

A.2.1 Constrained Delaunay triangulation and Delaunay mesher

In order to use an automatic Delaunay mesher we have to provide a closed loop of boundary segments. Within these boundaries points are inserted such that triangles that satisfy a maximum sidelength criterion and a minimum angle criterion are generated by the Delaunay mesher. The triangulation is forced to include constrained edges. Thus the triangulation is not necessarily Delaunay anymore, unless we allow additional points to be added on the constrained edges.

A.3 Elemental matrices and arrays of the discrete weak form for two-phase flow

The elemental matrices of an element e that are assembled into Equation 3.3.4 are given by:

- Body load vector $\mathbf{f}^e$

$$\int_{\Omega^e} \mathbf{N}^T \begin{bmatrix} C_s\rho_s & 0 & 0 & 0 & 0 \\ 0 & C_s\rho_s & 0 & 0 & 0 \\ 0 & 0 & C_f\rho_f & 0 & 0 \\ 0 & 0 & 0 & C_f\rho_f & 0 \\ 0 & 0 & 0 & 0 & 0 \end{bmatrix} \begin{bmatrix} b_x^s \\ b_y^s \\ b_x^f \\ b_y^f \\ 0 \end{bmatrix} d\Omega \quad (A.3.1)$$

- Nodal load vector $\mathbf{h}^e$

$$\int_{\Gamma^e} \mathbf{N}^T \begin{bmatrix} t_x^s \\ t_y^s \\ t_x^f \\ t_y^f \\ 0 \end{bmatrix} d\Gamma \quad (A.3.2)$$

- Mass matrix $\mathbf{M}^e$

$$\int_{\Omega^e} \mathbf{N}^T \begin{bmatrix} C_s\rho_s & 0 & 0 & 0 & 0 \\ 0 & C_s\rho_s & 0 & 0 & 0 \\ 0 & 0 & C_f\rho_f & 0 & 0 \\ 0 & 0 & 0 & C_f\rho_f & 0 \\ 0 & 0 & 0 & 0 & 0 \end{bmatrix} \mathbf{N} d\Omega \quad (A.3.3)$$

The lumped mass matrix used in this work is obtained by summing the rows of the above matrix.

- Gradient matrix $\mathbf{G}^e$

$$\int_{\Omega^e} \mathbf{B}^T \begin{bmatrix} C_s \\ C_s \\ 0 \\ C_f \\ C_f \\ 0 \end{bmatrix} \mathbf{N}^p d\Omega \qquad (A.3.4)$$

- Stiffness matrix $\mathbf{K}^e$

$$\int_{\Omega^e} \mathbf{B}^T \begin{bmatrix} C_s & 0 & 0 & 0 & 0 & 0 \\ 0 & C_s & 0 & 0 & 0 & 0 \\ 0 & 0 & C_s & 0 & 0 & 0 \\ 0 & 0 & 0 & C_f & 0 & 0 \\ 0 & 0 & 0 & 0 & C_f & 0 \\ 0 & 0 & 0 & 0 & 0 & C_f \end{bmatrix} \mathbf{DB} d\Omega \qquad (A.3.5)$$

- Momentum exchange matrix $\mathbf{V}^e$

$$\int_{\Omega^e} \mathbf{N}^T K_{drag} \begin{bmatrix} 1 & 0 & -1 & 0 & 0 \\ 0 & 1 & 0 & -1 & 0 \\ -1 & 0 & 1 & 0 & 0 \\ 0 & -1 & 0 & 1 & 0 \\ 0 & 0 & 0 & 0 & 0 \end{bmatrix} \mathbf{N} d\Omega \qquad (A.3.6)$$

- $\mathbf{K}^{\nabla \mathbf{C}^e}$

$$\int_{\Omega^e} \left\{ \mathbf{N}^T \begin{bmatrix} C_{s,x} & 0 & C_{s,y} & 0 & 0 & 0 \\ 0 & C_{s,x} & C_{s,x} & 0 & 0 & 0 \\ 0 & 0 & 0 & C_{f,x} & 0 & C_{f,y} \\ 0 & 0 & 0 & 0 & C_{f,y} & C_{f,x} \end{bmatrix} \mathbf{DB} + \mathbf{N}^T \begin{bmatrix} C_{s,x} \\ C_{s,y} \\ C_{f,x} \\ C_{f,y} \\ 0 \end{bmatrix} \mathbf{N}^p \right\} d\Omega \qquad (A.3.7)$$

- Stabilization matrix $\mathbf{S}^e$

$$\tau_e \int_{\Omega^e} (\nabla \mathbf{N}^p)^T \begin{bmatrix} C_s & 0 & 0 & 0 & 0 \\ 0 & C_s & 0 & 0 & 0 \\ 0 & 0 & C_f & 0 & 0 \\ 0 & 0 & 0 & C_f & 0 \\ 0 & 0 & 0 & 0 & 0 \end{bmatrix} \nabla \mathbf{N}^p d\Omega \qquad (A.3.8)$$

- Stabilization vector $\mathbf{f}_s^e$

$$\tau_e \int_{\Omega^e} (\nabla \mathbf{N}^p)^T \begin{bmatrix} C_s & 0 & 0 & 0 & 0 \\ 0 & C_s & 0 & 0 & 0 \\ 0 & 0 & C_f & 0 & 0 \\ 0 & 0 & 0 & C_f & 0 \\ 0 & 0 & 0 & 0 & 0 \end{bmatrix} \begin{bmatrix} b_x^s \\ b_y^s \\ b_x^f \\ b_y^f \\ 0 \end{bmatrix} d\Omega \qquad (A.3.9)$$

- N-matrix

$$\mathbf{N} = \begin{bmatrix} N^I & 0 & 0 & 0 & 0 & \dots \\ 0 & N^I & 0 & 0 & 0 & \dots \\ 0 & 0 & N^I & 0 & 0 & \dots \\ 0 & 0 & 0 & N^I & 0 & \dots \\ 0 & 0 & 0 & 0 & 0 & \dots \end{bmatrix} \qquad (A.3.10)$$

- $\mathbf{N}^p$-matrix

$$\mathbf{N}^p = \begin{bmatrix} 0 & 0 & 0 & 0 & N^I & \dots \end{bmatrix} \qquad (A.3.11)$$

- B-matrix

$$\mathbf{B} = \begin{bmatrix} N^I_{,x} & 0 & 0 & 0 & 0 & \dots \\ 0 & N^I_{,y} & 0 & 0 & 0 & \dots \\ N^I_{,y} & N^I_{,x} & 0 & 0 & 0 & \dots \\ 0 & 0 & N^I_{,x} & 0 & 0 & \dots \\ 0 & 0 & 0 & N^I_{,y} & 0 & \dots \\ 0 & 0 & N^I_{,y} & N^I_{,x} & 0 & \dots \end{bmatrix} \qquad (A.3.12)$$

- $\nabla \mathbf{N}^p$-matrix

$$\nabla \mathbf{N}^p = \begin{bmatrix} 0 & 0 & 0 & 0 & N^I_{,x} & \dots \\ 0 & 0 & 0 & 0 & N^I_{,y} & \dots \\ 0 & 0 & 0 & 0 & N^I_{,x} & \dots \\ 0 & 0 & 0 & 0 & N^I_{,y} & \dots \\ 0 & 0 & 0 & 0 & 0 & \dots \end{bmatrix} \qquad (A.3.13)$$

- Constitutive matrix D

$$D = \begin{bmatrix} \frac{4}{3}\mu_s & -\frac{2}{3}\mu_s & 0 & 0 & 0 & 0 \\ -\frac{2}{3}\mu_s & \frac{4}{3}\mu_s & 0 & 0 & 0 & 0 \\ 0 & 0 & \mu_s & 0 & 0 & 0 \\ 0 & 0 & 0 & \frac{4}{3}\mu_f & -\frac{2}{3}\mu_f & 0 \\ 0 & 0 & 0 & -\frac{2}{3}\mu_f & \frac{4}{3}\mu_f & 0 \\ 0 & 0 & 0 & 0 & 0 & \mu_f \end{bmatrix} \quad (A.3.14)$$

A.4 Computing volume fractions using mass conservation of each phase

Section 3.6 presents an overview of algorithmic approaches to compute volume fractions based on the updated nodal coordinates of the solid and the fluid phase. Another approach uses the equations of conservation of mass of the individual phases in their original form (Equation 3.2.10), formulated with respect to the material frame of reference. For constant mass density ρ_p we can write

$$\frac{DC_p}{Dt} + C_p \nabla \cdot \mathbf{v}_p = 0 \quad (A.4.1)$$

Combining mass conservation on a fixed reference frame allowed to obtain an equation of conservation of the volume of the mixture by eliminating the time derivatives of volume fractions. The idea is to introduce a time stepping algorithm to advance Equation A.4.1 in time. After iteration $i+1$ of time step $n+1$ we write

$$\frac{DC_p^{n+1}}{Dt} + C_p^{n+1} \nabla \cdot \mathbf{v}_{p,i+1}^{n+1} = 0 \quad (A.4.2)$$

Since $\mathbf{v}_{p,i+1}^{n+1}$ is known the above equation is an ordinary differential equation, which can be solved in each point of the domain separately. Using a generalized trapezoidal algorithm we can write the following finite difference formula:

$$\dot{C}_p^{n+1} = \frac{1}{\Delta t \gamma}(C_p^{n+1} - \tilde{C}_p^{n+1}) \quad (A.4.3)$$

$$\tilde{C}_p^{n+1} = C_p^n + \Delta t(1-\gamma)\dot{C}_p^n \quad (A.4.4)$$

where $\dot{C}_p^{n+1} = \frac{DC_p^{n+1}}{Dt}$. After iteration $i+1$ the volume fraction of phase p can be computed according to

$$C_p^{n+1} = \frac{1}{1 + \Delta t \gamma \nabla \cdot \mathbf{v}_{p,i+1}^{n+1}} \tilde{C}_p^{n+1} \quad (A.4.5)$$

We propose two different approaches:

- **Strong form**: Equation A.4.5 is solved at each Gauss point of the deformed mesh of phase p. In this case the divergence of the velocity can easily be obtained using the shape function derivatives.

- **Weak form**: A weak form of Equation A.4.5 can be obtained by multiplication with a test function. This results in a system of equations of the form

$$\mathbf{A} C_p^{n+1} = \mathbf{F}(\mathbf{v}_{p,i+1}^{n+1}, \tilde{C}_p^{m+1}) \tag{A.4.6}$$

where C_p^{n+1} is a vector of nodal values of volume fraction. $\mathbf{A}$ has the form of a mass matrix. Using a lumped matrix $\mathbf{A}$ allows to compute the solution without having to solve a system of equations.

In both approaches the values of C_p^{n+1} have to be mapped onto the new mesh. From here on the procedure is the same as previously, correcting for mass conservation of both phases according to point 3 in Table 3.4.

Bibliography

[1] M.J. Andrews and P.J. O'Rourke. The multiphase particle-in-cell (mp-pic) method for dense particulate flows. *International Journal of Multiphase Flow*, 22(2):379–402, 1996.

[2] I. Babuška. Error bounds for finite-element method. *Numerische Mathematik*, 16:322–333, 1971.

[3] T. Belytschko, Y. Krongauz, D. Organ, M. Fleming, and P. Krysl. Meshless methods: An overview and recent developments. *Computer Methods in Applied Mechanics and Engineering*, 139(1-4):3–47, 1996.

[4] T. Belytschko, W. K. Liu, and B. Moran. *Nonlinear Finite Elements for Continua and Structures*. Wiley, Chichester, 2000.

[5] H. Braess and P. Wriggers. Arbitrary lagrangian eulerian finite element analysis of free surface flow. *Computer Methods in Applied Mechanics and Engineering*, 190(1-2):95–109, 2000.

[6] F. Brezzi. On the existence, uniqueness and approximation of saddle-point problems arising from lagrangian multipliers. *Revue Française d'Automatique Inf. Rech. Oper.*, 8:129–151, 1974.

[7] F. Brezzi and J. Pitkäranta. On the stabilization of finite element approximations of the stokes equations. In Wolfgang Hackbusch, editor, *Efficient Solutions of Elliptic Systems*, volume 10 of *Notes on Numerical Fluid Mechanics*, pages 11–19, Braunschweig, Germany, January 1984. Vieweg.

[8] A. Caboussat. *Analysis and numerical simulation of free surface flows*. PhD thesis, Ecole Polytechnique Federale de Lausanne, 2003. Thesis no. 2893.

[9] CGAL, Computational Geometry Algorithms Library. http://www.cgal.org.

[10] C. Chen. General solutions for viscoplastic debris flow. *Journal of Hydraulic Engineering*, 114(3):259–282, 1988.

[11] C. Chen. Generalized viscoplastic modeling of debris flow. *Journal of Hydraulic Engineering*, 114(3):237–258, 1988.

[12] R. Codina. Pressure stability in fractional step finite element methods for incompressible flows. *Journal of Computational Physics*, 170(1):112–140, 2001.

[13] S. Commend and Th. Zimmermann. Object-oriented nonlinear finite element programming: a primer. *Advances in Engineering Software*, 32:611–628, 2001.

[14] E. Cueto, M. Doblaré, and L. Gracia. Imposing essential boundary conditions in the natural element method by means of density-scaled α-shapes. *International Journal for Numerical Methods in Engineering*, 49:519–546, 2000.

[15] V. de la Cruz and T. J. T. Spanos. Mobilization of oil ganglia. *AIChE Journal*, 29(5):854–858, 1983.

[16] M. K. Denham, P. Briard, and M. A. Patrick. A directionally-sensitive laser anemometer for velocity measurements in highly turbulent flows. *Journal of Physics E: Scientific Instruments*, 8:681–683, 1975.

[17] M. K. Denham and M. A. Patrick. Laminar flow over a downstream-facing step in a two-dimensional flow channel. *Transactions of the Institution of Chemical Engineers*, 52:361–367, 1974.

[18] R. P. Denlinger and R. M. Iverson. Flow of variably fluidized granular masses across 3-d terrain: 2. numerical predictions and experimental tests. *Journal of Geophysical Research*, 106(B1):553–566, 2001.

[19] J. Dolbow and T. Belytschko. Volumetric locking in the element free galerkin method. *International Journal for Numerical Methods in Engineering*, 46(6):925–942, 1999.

[20] D. Eyheramendy. New tracks for future computational platforms for engineering applications. In Th. Zimmermann and A. Truty, editors, *Numerics in Geotechnics*, pages 1–15. Elmepress Int., 2006.

[21] D. González, E. Cueto, F. Chinesta, and M. Doblaré. A natural element updated lagrangian strategy for free-surface fluid dynamics. *Journal of Computational Physics*, 223(1):127–150, 2007.

[22] D. González, M. A. Cueto, and M. Doblaré. Volumetric locking in natural neighbour galerkin methods. *International Journal for Numerical Methods in Engineering*, 61(4):611–632, 2004.

[23] J.L. Guermond, P. Minev, and J. Shen. An overview of projection methods for incompressible flows. *Computer Methods in Applied Mechanics and Engineering*, 195(44–47):6011–6045, 2006.

[24] T. J. R. Hughes. *The Finite Element Method*. Dover Publications, Mineola (NY), 2000.

[25] T. J. R. Hughes, L. P. Franca, and M. Balestra. A new finite-element formulation for computational fluid-dynamics: V. circumventing the Babuska-Brezzi condition - a stable petrov-galerkin formulation of the stokes problem accomodating equal-order interpolations. *Computer Methods in Applied Mechanics and Engineering*, 59(1):85–99, November 1986.

[26] T. J. R. Hughes, W. K. Liu, and Th. Zimmermann. Lagrangian-eulerian finite element formulation for incompressible viscous flows. *Computer Methods in Applied Mech. Engineering*, 29:329–349, 1981.

[27] T. J. R. Hughes, K. S. Pister, and R. L. Taylor. Implicit-explicit finite elements in nonlinear transient analysis. *Computer Methods in Applied Mechanics and Engineering*, 17/18:159–182, 1979.

[28] T.J.R. Hughes, L.P. Franca, and G.M. Hulbert. A new finite element formulation for computational fluid dynamics: Viii. the galerkin/least-squares method for advective-diffusive equations. *Computer Methods in Applied Mechanics and Engineering*, 73(2):173–189, 1989.

[29] K. Hutter. Debris and mudflows: Are we asking the correct questions? what are the deficits? *Mitteilungen der Versuchsanstalt fur Wasserbau, Hydrologie und Glaziologie an der Eidgenössischen Technischen Hochschule Zürich*, 190:91–105, 2005.

[30] K. Hutter, B. Svendsen, and D. Rickenmann. Debris flow modeling: A review. *Continuum Mechanics and Thermodynamics*, 8(1):1–35, 1996.

[31] S. R. Idelsohn, E. Oñate, and F. Del Pin. The particle finite element method: a powerful tool to solve incompressible flows with free-surfaces and breaking waves. *International Journal for Numerical Methods in Engineering*, 61:964–989, 2004.

[32] R. M. Iverson and R. P. Denlinger. Flow of variably fluidized granular masses across 3-d terrain: 1. coulomb mixture theory. *Journal of Geophysical Research*, 106(B1):537–552, 2001.

[33] Nikolay I. Kolev. *Multiphase flow dynamics*. Springer, Berlin, 2nd edition edition, 2005.

[34] D. Laigle, P. Lachamp, and M. Naaim. Sph-based numerical investigation of mudflow and other complex fluid flow interactions with structures. *Computational Geosciences*, 11(4):297–306, 2007.

[35] I. Malcevic and O. Ghattas. Dynamic-mesh finite element method for lagrangian computational fluid dynamics. *Finite Elements in Analysis and Design*, 38(10):965–982, 2002.

[36] M. A. Martínez, E. Cueto, I. Alfaro, M. Doblaré, and F. Chinesta. Updated lagrangian free surface flow simulations with natural neighbour galerkin methods. *International Journal for Numerical Methods in Engineering*, 60(13):2105–2129, 2004.

[37] F. T McKenna. *Object Oriented Finite Element Programming Frameworks for Analysis Algorithms and Parallel Computing*. Phd thesis, University of California at Berkeley, 1990.

[38] A. Munjiza and K. R. F. Andrews. Nbs contact detection algorithm for bodies of similar size. *International Journal for Numerical Methods in Engineering*, 43(1):131–149, 1998.

[39] W. L. Oberkampf, T. G. Trucano, and Ch. Hirsch. Verification, validation, and predictive capability in computational engineering and physics. *Applied Mechanics Reviews*, 57(1–6):345–384, 2004.

[40] J.S. O'Brien, P.Y. Julien, and W.T. Fullerton. Two-dimensional water flood and mudflow simulation. *Journal of Hydraulic Engineering - ASCE*, 119(2):244–261, 1993.

[41] E. Oñate, S. Idelsohn, O. C. Zienkiewicz, and R. L. Taylor. A finite point method in computational mechanics. applications to convective transport and fluid flow. *International Journal for Numerical Methods in Engineering*, 39(22):3839–3866, 1996.

[42] E. Oñate, J. Rojek, R. L. Taylor, and O. C. Zienkiewicz. Finite calculus formulation for incompressible solids using linear triangles and thetrahedra. *International Journal for Numerical Methods in Engineering*, 59:1473–1500, 2004.

[43] C. Potter. Ecoulements en surface libre. Internal report LSC, EPFL, Lausanne.

[44] M.A. Puso, J.S. Chen, E. Zywicz, and W. Elmer. Meshfree and finite element nodal integration methods. *International Journal for Numerical Methods in Engineering*, 74(3):416–446, 2008.

[45] Radovitzky R. and Ortiz M. Lagrangian finite element analysis of newtonian fluid flows. *International Journal for Numerical Methods in Engineering, 43 (4), pp. 607-619*, 43(4):607–619, 1998.

[46] S. B. Savage and K. Hutter. The motion of a finite mass of granular material down a rough incline. *Journal of Fluid Mechanics*, 199:177–215, 1989.

[47] S. Shao and E.Y.M. Lo. Incompressible sph method for simulating newtonian and non-newtonian flows with a free surface. *Advances in Water Resources 26 (7), pp. 787-800*, 26(7):787–800, 2003.

[48] S. L. Soo. *Multiphase Fluid Dynamics*. Science Press, Beijing, China, 1990.

[49] N. Sukumar. *The natural element method in solid mechanics*. Phd thesis, Northwestern University, Evanston, IL, U.S.A., June 1998.

[50] N. Sukumar, B. Moran, and T. Belytschko. The natural element method in solid mechanics. *International Journal for Numerical Methods in Engineering*, 43:839–887, 1998.

[51] I. E. Sutherland, R. F. Sproull, and R. A. Schumacker. A characterization of ten hidden-surface algorithms. *ACM Computing Surveys*, 6(1):1–55, 1974.

[52] B. Svendsen and K. Hutter. On the thermodynamics of a mixture of isotropic materials with constraints. *International Journal of Engineering Science*, 33(14):2021–2054, 1995.

[53] Tamotsu Takahashi. *Debris flow*. IAHR Monograph Series. A. A. Balkema, Rotterdam, Brookfield, 1991.

[54] T. E. Tezduyar, M. Behr, and J. Liou. A new strategy for finite element computations involving moving boundaries and interfaces-the deforming-spatial-domain/space-time procedure: I. the concept and the preliminary numerical tests. *Computer Methods in Applied Mechanics and Engineering*, 94(3):339–351, 1992.

SVH Südwestdeutscher Verlag für Hochschulschriften

Wissenschaftlicher Buchverlag bietet
kostenfreie
Publikation
von
Dissertationen und Habilitationen

Sie verfügen über eine wissenschaftliche Abschlußarbeit zu aktuellen oder zeitlosen Fragestellungen, die hohen inhaltlichen und formalen Anspruchen genügt, und haben **Interesse an einer honorarvergüteten Publikation?**

Dann senden Sie bitte erste Informationen über Ihre Arbeit per Email an: info@svh-verlag.de.

Unser Außenlektorat meldet sich umgehend bei Ihnen.

Südwestdeutscher Verlag für Hochschulschriften
Aktiengesellschaft & Co. KG

Dudweiler Landstr. 99
D – 66123 Saarbrücken
www.svh-verlag.de

Printed by Books on Demand GmbH, Norderstedt / Germany